KU-740-812

Non fiction Gift Aid
£

031140 002821

POETRY

Bm

24 / 24

Thatcherism and the Fall of Coal

Thatcherism and the Fall of Coal

M. J. PARKER

Published by the Oxford University Press
for the Oxford Institute for Energy Studies
2000

Oxford University Press, Great Clarendon Street, Oxford OX2 6DP

Oxford University Press is a department of the University of Oxford.
It furthers the University's objective of excellence in research, scholarship
and education by publishing worldwide in

Oxford New York

Athens Auckland Bangkok Bogotá Buenos Aires Calcutta
Cape Town Chennai Dar es Salaam Delhi Florence Hong Kong
Istanbul Karachi Kuala Lumpur Madrid Melbourne
Mexico City Mumbai Nairobi Paris São Paulo
Singapore Taipei Tokyo Toronto Warsaw
with associated companies in Berlin Ibadan

Oxford is a registered trade mark of Oxford University Press
in the UK and in certain other countries

Published in the United States
by Oxford University Press Inc. New York

© Oxford Institute for Energy Studies 2000

First published 2000

All rights reserved. No part of this publication may be reproduced, stored in a retrieval
system, or transmitted, in any form or by any means, without the prior permission in
writing of the Oxford Institute for Energy Studies, or as expressly permitted by law, or
under terms agreed with the appropriate reprographics rights organization. Enquiries
concerning reproduction outside the scope of the above should be sent to
Oxford Institute for Energy Studies, 57 Woodstock Road, Oxford OX2 6FA.

You must not circulate this book in any other binding or cover
and you must impose this same condition on any acquirer

British Library Cataloguing in Publication Data
Data Available

Library of Congress Cataloguing in Publication Data
Data Applied for

ISBN 0-19-730023-5

Cover designed by Oxford Designers and Illustrators
Typeset by Philip Armstrong, Sheffield
Printed by Biddles, Guildford

To my wife Penny

ACKNOWLEDGEMENTS

I am indebted to Dr Nigel Evans and Professor Alexander Kemp, who were instrumental in persuading me to undertake this study, and to Professor John Surrey for his help in turning my initial ideas into a workable framework.

Thanks are due also to those who have provided comment and advice on successive drafts, and in particular to Mike Atkinson, Anthony Baker, David Brandrick, Neil Clarke, Tony Cooper, Malcolm Edwards, Dave Feickert, Dieter Helm, Charles Henderson, Eddy Hindmarsh, Andrew Horsler, Philip Hutchinson, David Jones, Malcolm Keay, David Merrick, Professor Gerald Manners, Ned Smith, and Jim Watson.

I am also grateful to the British Coal Corporation (which continued as a legal entity after the privatisation of its mining assets in 1994) for permission to consult the administrative papers and records of the National Coal Board/British Coal Corporation; to Paul Horsnell of the Oxford Institute for Energy Studies for his help and encouragement; and to Charlotte Huggett at SPRU for organising supporting typing and secretarial facilities.

My gratitude to those who have helped makes it the more important to emphasise that any errors of fact or mistaken judgements are solely my responsibility.

M. J. Parker

CONTENTS

LIST OF TABLES

FOREWORD

By the early 1980s the British coal industry appeared to have achieved stability after a long period of decline. In the six years to 1982, total annual output averaged over 120m. tonnes with deep-mined output over 100m. tonnes a year. In 1982, colliery manpower still exceeded 200,000 at nearly 200 pits. However, Professor William Ashworth, concluding his magisterial account of the nationalised coal industry down to 1982, wrote that

> there were many great uncertainties in a rapidly changing and depressed economic environment, there was the prospect of intense competition in energy supply for years ahead, and there were various uncured weaknesses and inefficiencies in the coal industry itself. For all these reasons the phase of coal industry history which came down to the early nineteen-eighties had to have, as its closing symbol, a large question mark.
> (W Ashworth: *The History of the British Coal Industry*: Volume 5: *1946–1982 The Nationalised Industry*, p.670)

Ashworth's conclusion proved to be both prescient and understated. By 1999 deep-mined output had fallen to little more than 20m. tonnes with the prospect of further decline, and the colliery labour force had fallen to well below 10,000. This process of 'decline and fall' has been accompanied by much controversy. In terms of explanation, the extremes of the argument can be characterised by three opposing views: first, that the crucial factor was the political agenda of successive Conservative governments from 1979 through to the privatisation of British Coal, designed irreversibly to break the power of the National Union of Mineworkers (NUM), and thereby weaken the trade union movement and the socialist cause; or second, that the coal industry was run down by powerful market forces which exposed its fundamental economic weakness; or third, that the industry (both management and miners) was the engine of its own destruction.

However, the simplifications inherent in these opposing views are more likely to lead to the generation of myths than to an understanding of the realities.

The whole complex story cannot be understood except in terms of analysis which charts the *interactions* of government policy, market forces, and the industry's own performance, over the whole period from 1979 down to the end of the century.

As the author was for many years a close and not disinterested observer of events, he is very conscious of the dangers attending a study of this kind: dangers of political prejudice, and of imposing upon the story a greater degree of coherence (particularly in terms of the motives and intentions of the major players) than would be justified without the benefit of hindsight. In particular, where he has sought to redress the generally low public esteem in which the National Coal Board/British Coal Corporation was held, he has tried to do so by objective reference to the evidence. It will be for the reader to judge whether the dangers of partiality have been avoided.

CHAPTER I
THE ARRIVAL OF MRS THATCHER:
CONFRONTATION POSTPONED

OPEC Oil Crisis: Would it Help Coal?

Margaret Thatcher's Conservative government came to power in May 1979 in the middle of the second OPEC 'oil crisis', during which the price of crude oil was to increase from $13 to $34 per barrel. The Shah of Iran had gone into exile, Iranian oil exports had ceased, and world oil markets were in turmoil. Although Western governments had taken collective action following the first OPEC 'oil shock' in 1973/4 (for example, by the creation of the International Energy Agency), in 1979 there was renewed alarm about security of energy supplies, and a perceived need to use all possible means to reduce dependence on oil, for both economic and strategic reasons.

These events seemed to provide a very favourable strategic framework for the UK coal industry. Developments in the UK energy market appeared to confirm this. Since the first OPEC crisis of 1973/4, coal's share of the UK primary energy market had stabilised at around 36 per cent; and in 1979/80 NCB output was the highest for four years and UK coal consumption the highest for seven years. Above all, UK power stations were using record amounts of coal, which made up three-quarters of all the fuel used for electricity generation, and UK coal prices were well below the price of fuel oil. No significant alternatives to coal were available. The AGR nuclear programme was still subject to long construction delays, and there was continuing controversy on the choice of future reactors. Natural gas was at that time regarded as a 'premium' fuel, too valuable to be used in power station boilers. Future prices and supplies of internationally-traded power station coal were then very uncertain. For all these reasons, at the time it appeared inevitable that UK coal's share of the power generation fuel market would continue at a high level for the foreseeable future.

The new Conservative government was drawn into the international enthusiasm for coal as a means of reducing oil dependence, and was party to the final Communiqué of the Economic Summit held in Tokyo in June 1979, under which Heads of Governments of the seven leading industrial nations (G7) pledged their countries 'to increase as far as possible coal use, production and trade without damage to the environment', and also to 'maintain by measures which do not obstruct coal imports, those levels of domestic production which are desirable for reasons of energy, regional and social policy'. Almost the first act of the new Secretary of State for Energy (David Howell) was to agree in May 1979 to the International Energy Agency's 'Principles for IEA Action on Coal', which affirmed the importance of coal in energy policy, and stated that 'beyond 1985 coal could provide a substantially greater contribution to the energy needs of IEA countries. This depends on the adoption by governments now of appropriate coal policies which stimulate capital investment on a scale commensurate with the long-term potential of this energy source' (quoted in *NCB Annual Report 1979/80*, p.5). Further, the Department of Energy's evidence (July 1979) to the Vale of Belvoir Inquiry (at which the NCB was seeking planning permission for three new mines) stated that 'it is essential the development of efficient, modern capacity should proceed to provide the basis for future expansion of output'. In April 1980, the UK government provided members (which included the NCB Chairman, Derek Ezra) for the IEA's Coal Industry Advisory Board (CIAB), set up as a result of the Tokyo summit of 1979.

At the G7 Economic Summit held in Venice in June 1980, in which Mrs Thatcher participated, the governments collectively pledged themselves to break the links between economic growth and oil consumption, through the promotion of energy savings and the development of alternative sources of energy, including coal. Yet Margaret Thatcher was very reluctant to use these circumstances to endorse the geopolitical or strategic case for UK coal. She had taken some persuasion to agree the pro-coal sentiments in the Venice communiqué, and insisted that, whatever the *collective* agreement of G7 governments, this did not commit the UK government to expand its own coal industry. To understand this, we need to look back to the events in 1973/4 which led to the 'Plan for Coal', a major capital programme designed to modernise the coal industry, and prevent its decline, following a prolonged period of low investment.

'Plan for Coal' in Retrospect

The international energy situation in 1979 was not dissimilar to that which had faced the Heath government in the early 1970s. At that time, the NCB argued strongly that, in view of the deteriorating position in the Middle East and the increasing power of OPEC, there was a strong economic and security case for UK coal. Although in early 1972 the industry had suffered its first national strike since 1926 (which had been resolved only by the 'Wilberforce' enquiry effectively granting the whole of the NUM's very large wage claims), the NCB analysis was widely shared, both by senior figures in the oil industry, and in government circles.

For several years, officials of the NCB (including the author) had been involved in informal discussions with the Central Policy Review Staff (CPRS), which was headed by Lord Rothschild and reported to the Prime Minister (Mr Heath). Encouragement for the NCB position also came from the Department of Trade and Industry (DTI) and the Secretary of State, Peter Walker, who at a meeting in May 1973 asked the Board Chairman, Derek Ezra, how soon a start could be made on new mines. The outcome of these informal talks was the submission by the NCB of a document, 'The Economic Case for Coal', as the first stage of preparation of a 'Plan for Coal', to be submitted to the government in the autumn. The main points of the document were: 'The economic case for coal rests principally on the expectation of coal improving its competitive position in relation to oil'. Depending on the extent and timing of expected oil price increases, it was highly probable that average coal costs would be below the price of fuel oil before 1980, particularly in the central coalfields (Yorkshire and the Midlands), where the opportunities for further investment were concentrated.

Due to the low level of investment in major projects, deep-mined output would fall by some 45m. tonnes by the mid-1980s, with the losses being most marked in the central coalfields.

The NCB would need to draw up a ten-year development plan which would set out the opportunities for economic investment at new mines and the best of existing collieries, the scope for further productivity improvements, the financing implications of a major investment programme, and the manpower problems arising from likely contraction at high cost collieries in the 'peripheral' coalfields (particularly Scotland, the North East and South Wales).

The process of preparation continued during the summer of 1973. A 'Coal Prospects Committee' was set up to co-ordinate work of the NCB and DTI, including the possibility of facilitating any public inquiries that might be required for new mines. In the event, the outbreak of the Arab/Israeli 'Yom Kippur' war precipitated the first OPEC 'oil crisis' in October 1973. Coincidentally, the first (un-published) version of 'Plan for Coal' was submitted to the government on 15 October 1973, under circumstances which heightened the political urgency of consideration of the coal industry's future. A special meeting on the 'oil crisis' was held at Sunningdale on 18/19 October 1973, chaired by Industry Minister Tom Boardman, and attended by heads of oil companies and nationalised energy industries, and senior civil servants. In the absence of the Board Chairman through illness, the NCB was represented by the Deputy Chairman (W.V. Sheppard) and the author, who emphasised that the main contribution that coal could make was in substitution for fuel oil in power stations, but that, given that the UK coal industry was an old extractive industry, even a major investment programme could not be expected to do more than prevent decline. On that basis, however, such a programme would make a worthwhile contribution to reducing dependence on oil.

By December, crude oil prices were three times the pre-October levels, thereby creating a very large potential competitive headroom for coal. Following a presentation to the Secretary of State in December, the NCB prepared a revised version of 'Plan for Coal', which was finally submitted to the government in February 1974, which took into account representations from the DTI to see whether planned output in 1985 might be further increased, and whether the programme might be accelerated 'so as to produce a greater impact over the next five critical years' (Private communication 6.12.73). The February 1974 version of 'Plan for Coal' (which was never published) envisaged an output and demand profile, given a major investment programme, as shown in Table 1.1.

The final recommendations nevertheless recognised that 'plans must be kept flexible, and that a rigid long-term output target would be unrealistic', and that 'individual schemes will have to be considered on

Table 1.1: 'Plan for Coal' Output and Demand Profile. Million Tonnes

	1972/3*	1980	1985
Output			
Deep mines	127	120	120
Opencast etc	11	15	15
	138	135	135
Demand			
Power stations	69	83	88
Other markets	61	52	47
	130	135	135

* Adjusted to 52 week basis (1972/3 was 53 week year)

their merits', using full DCF financial appraisal. Specifically, the NCB recommended that the government should agree in principle that

(i) the Board should undertake a 'Plan for Coal' based on an investment programme (of major projects) ... designed to yield over 40m. tons of capacity by the mid-1980s, primarily of low-cost general purpose (steam) coal in the Midlands and Yorkshire;

(ii) subject to satisfactory completion of proving (exploration) and detailed cost estimates, this investment programme should include the development of the reserves at Selby and other new mines;

(iii) appropriate measures be taken to make available to the Board the funds necessary to carry the 'Plan' through, including possible alternative methods of financing to the extent to which borrowings will exceed self-generated funds;

(iv) any other adjustment which might be necessary to bring supply into line with demand, after taking account of reasonable levels of stocks, should be achieved by carefully controlled closure of the higher-cost capacity rather than restriction of investment in low-cost capacity.

(v) opencast coal output should be maximised, and the opening of new sites facilitated;

(vi) early consideration should be given to plans for new coal-fired (generating) capacity to be commissioned by the mid-1980s;

(vii) a government-sponsored programme of increased research should be drawn up.

There is little doubt that, had it continued in office, the Heath government would have implemented a policy along these lines. However, by the time the February 1974 version of 'Plan for Coal' had been submitted, other events had assumed centre stage. In July 1973, the NUM had formally submitted a wage claim representing increases of 31 per cent, which put the union on collision course with the government's counter-inflationary policy, which the NCB, as a nationalised industry, was obliged to defend. The result was an NUM overtime ban from 12 November 1973 and national strike from 9 February 1974, which ran in parallel with the General Election campaign called by Edward Heath on the issue of 'who governs Britain'. The strike did not end until 9 March, by which time a new Labour government had taken office and had immediately agreed a wage settlement broadly in line with the NUM's claim. The President of the NUM (Joe Gormley) and his more moderate supporters on the NUM Executive, had had no wish to use the coal strike to force a change of government. But Gormley was finding it increasingly difficult to contain the influence of the militants, led by the Communist Vice-President (Mick McGahey) and the young Arthur Scargill, who made no secret that this was their

aim. (Edward Heath: *The Course of My Life,* p.505). While there is room
for argument as to whether Heath could have avoided an election, and
therefore whether the fall of his government had been the result of
political miscalculation, after the event the power of the NUM was
generally regarded as the decisive factor in overthrowing the
Conservative government.

It is against this background that further discussion of 'Plan for
Coal' was resumed, but, on the initiative of the new Secretary of State
(Eric Varley) on a 'tripartite' basis, with representatives of Government,
the Board, and the Mining Unions. The process, which was named
the 'Coal Industry Examination', began in April 1974 and published
its 'Interim' Report in June and 'Final' Report in October. Comparisons can be made between the two Coal Industry Examination (CIE)
Reports and the (unpublished) February version of 'Plan for Coal',
which had been submitted to the previous Conservative government,
in particular:-

- Although the CIE reports endorsed the 'Plan for Coal' programme
  of investment to create 42m. tonnes of new and replacement capacity
  by the mid 1980s, at a cost of £600m at 1974 prices, there was
  greater emphasis on the need to expand total deep-mined output to
  meet a higher possible view of market demand of up to 150m.
  tonnes by 1985 (Interim Report paras. 21 and 26).
- Although the Interim Report stated (para. 8 of Foreword) that 'for
  mineworkers ... real rewards can advance to the extent that they
  reflect higher efficiency and higher output per man,' (assumed in
  the Plan to increase by 4 per cent p.a.), in the Final Report (para
  16) this clear link between productivity and wages was lost. Instead
  it is stated that 'operating at maximum efficiency should enable the
  rewards of those employed in the industry to be sufficient to secure
  the manpower and skills necessary over the years to achieve that
  (target) production' – although 'we hope ... that the Board and the
  NUM can work out as soon as possible an effective and viable
  incentive scheme' (para. 57).
- The Final Report stated (para 10) that while the industry should
  aim to 'supply at a cost which can meet the long-run competition
  from other sources of energy ... it was the Government's intention
  that the industry should not be at the mercy of short-term
  fluctuations in the prices of competing fuels ... The Government
  will if necessary be prepared to help if temporary problems of this
  kind develop.' Moreover, (para 78) the new financial framework for
  the industry would recognise 'the need to take appropriate action if

other public policies prevent commercial pricing or impose exceptional burdens on the Board'.

- There was ambivalence on the issue of 'economic' closures. The Interim Report (para 23) merely stated that 'over the period up to 1985 it appears that a broad average of some 3–4m. tons capacity a year is likely to be lost mainly through exhaustion of mines and possibly also through exceptional mining difficulties'. The Final Report was even less specific: 'Inevitably some pits will have to close as their useful economic reserves of coal are depleted' (para. 27).

The two Reports of the Coal Industry Examination in 1974, which were effectively the vehicle for the new Labour government's endorsement of the NCB's 'Plan for Coal', bear all the marks of circumspect drafting arising from the public nature of the deliberations and the presence of the Mining Unions. When compared with the original 'Plan for Coal' submitted to the Conservative government, the final Plan had acquired, in various subtle ways, a greater emphasis on volume targets, with less economic and financial rigour. For the Mining Unions at least, the endorsement of 'Plan for Coal' represented an unconditional political commitment to the expansion of the industry.

There were also real problems in implementation of the Plan. Not only were the capital costs of the investment programme substantially greater in real terms than originally projected, the performance on productivity and operating costs was poor. Notwithstanding the 'Plan for Coal' assumption of 4 per cent per annum increase in output per man, in fact, it was not until 1981/2 that labour productivity exceeded the level seen in 1972/3; and few high-cost collieries were closed. Due to these and other factors, average colliery operating costs rose by almost 50 per cent in real terms between 1972/3 and 1979/80, storing up serious economic problems for the industry in the 1980s.

Although 'Plan for Coal' had been initiated before the first OPEC crisis of 1973/4 with the active encouragement of Mr Heath's Conservative government, and at a time when there was a good strategic and economic case for halting the decline of the UK industry, this did nothing to commend the coal industry to Mrs Thatcher when she became Prime Minister during the second OPEC crisis of 1979/80. In Conservative minds (particularly those of a right-wing persuasion), both the accession of the Labour government in March 1974, and its rapid endorsement of 'Plan for Coal', had been due to the victory of the Miners' Strike, with a result that the fortunes of the coal industry had become closely identified with the Labour Party and the interests of the NUM. Thus, the political circumstances under which a coal policy based on 'Plan for Coal' had been launched, contributed decisively to the political polarisation which made it almost impossible to establish a bipartisan approach to the nationalised coal industry's affairs.

Margaret Thatcher's New Policies for UK Coal

Although the second OPEC crisis seemed to many to show the wisdom of following a policy of long-term support for the UK coal industry on strategic energy policy grounds, such views were quite contrary to those of Margaret Thatcher, who was later to write: 'By the 1970's the coal mining industry had come to symbolize everything that was wrong with Britain' (Margaret Thatcher: *The Downing Street Years*, p.340). The NCB was a nationalised industry founded on monopoly. Above all, this monopoly working with the very large increases in OPEC oil prices had conferred great power on the NUM, the praetorian guard of the Labour movement, whose national strikes had caused an earlier Conservative government to submit to its demands in 1972 and had effectively been the occasion of the fall of that government in 1974. In his memoirs, Nicholas Ridley says that in 1978, when he was preparing his report on future policy on privatisation, he was asked to include a confidential annex on how a future Conservative government could defeat the NUM if faced with another confrontation. He added that 'in my view (Mrs Thatcher) always knew, even as far back as 1978, that she would face a pitched battle mounted by Arthur Scargill' (Nicholas Ridley: *My Style of Government*, p.67). Although Ridley wrote with the benefit of hindsight on this point, there is certainly no doubting that from the outset Margaret Thatcher had a great aversion to the perceived power of the NUM to 'hold the country to ransom' (which was seen as an extreme example of the wider problem of trade union power); and an equally strong desire not to fall victim to the NUM as Edward Heath had done. There was also a strong wish to change what was seen as the inefficient and non-commercial way in which the NCB was operated. This was articulated by Nigel Lawson who, writing to Geoffrey Howe in early 1981, stated that 'our original aim was to build a successful, profitable coal industry independent of government subsidies; to de-monopolise it and ultimately open it to private enterprise' (Nigel Lawson: *The View from Number 11*, p.142).

There is no doubt that these Thatcherite sentiments, strongly held over a long period, represented the fundamental aims of the

Conservative government's policy towards the coal industry, to be implemented as circumstances changed and opportunities for particular measures presented themselves. These fundamental aims on coal policy had a wider political significance, not only because they derived from the general Thatcherite aversion to trade union power and public ownership, but also because success with coal policy would have a considerable bearing on the success of the Thatcherite enterprise as a whole. In that sense, the aims of coal policy could be said to represent a 'political agenda'.

Nevertheless, there is no evidence that the new government came into power in 1979 with a clearly thought-out long-term master plan designed to effect radical change in the coal industry. There was no serious consideration of measures designed to 'break the power of the NUM': indeed to have done so at a time when the NUM retained the ability to disrupt electricity supplies would have been highly imprudent. And in 1979/80, the new government did not single out the coal industry for special financial treatment. The review of the NCB's financial framework was seen as part of the government's macro-economic policy, with its emphasis on the control of the money supply and reducing the Public Sector Borrowing Requirement (PSBR): an approach which applied to all nationalised industries, not just the NCB.

The coal industry's contribution to the PSBR was represented by its External Financing Requirement (EFR), made up of two main elements: government grants and net borrowing to finance investment and working capital, whether or not such borrowings were from the National Loans Fund (NLF). For control purposes, performance was monitored against External Financing Limits (EFL), which became known colloquially as 'cash limits'. By 1979/80, the NCB's financing requirements were on a sharply rising trend, and were outstripping the rise in mining investment (See Table 1.2).

Table 1.2: NCB External Financing Requirements: 1977/8 to 1980/1. £million

	1977/8	1978/9	1979/80	1980/1
Government grants	75	172	251	254
Additional loans	240	415	393	586
Total external finance required	315	587	644	840
Capital expenditure on mining	334	454	617	736
Financing of stocks	79	118	4	327

Source: *NCB Annual Reports*

Although the new Conservative government showed no sign of wishing to cut back the high level of capital expenditure on major projects inherited under the 'Plan for Coal' programme, nevertheless they wanted the NCB to play its part in reducing the PSBR by achieving a greater degree of self-financing of investment. As early as August 1979, at the request of government, the NCB agreed to a new Financial Strategy for the industry covering the years 1979/80 to 1983/4 under which the NCB would achieve viability, that is achieve at least an overall balance on revenue account after meeting interest charges by 1983/4, subject only to the receipt of certain 'social cost' grants; and the NCB's EFLs were to be substantially reduced in real terms beyond 1980/1, requiring an increasing proportion of the industry's large investment programme to be met from operating profits. This new financial framework was given statutory force in the Coal Industry Act 1980, which provided (inter alia) for an extension of the Board's borrowing powers and for further loans to be provided by Parliament; and a 'deficit' grant to be made to reduce or eliminate any deficit on the NCB consolidated profit and loss account. But the power to make regional grants to the Board was repealed; and an overall limit of £525m. was placed on operating and deficit grants for the years ending March 1980 to March 1983 (although this could be raised to £590m. with the approval of Parliament).

In April 1980, the Parliamentary Under Secretary of State, speaking on the Coal Industry Bill, said that the government had concluded a major review of the policy for coal, and that

> Our policy is based on two principles. The first is that, if the coal industry is competitive and based on efficient, high productivity capacity, it will have an essential and increasing part to play in meeting our future needs for energy. ... There will be an expanding market for coal at the right price ...

> The second principle of our policy is that the industry must be put on a sound commercial and financial footing for the future. During a difficult period of change, the industry has come to need substantial government grants. The government is naturally concerned to reduce the degree of financial support which it now gives. But the industry, too, will want to free itself progressively from dependence on government funds and to make itself fully viable and competitive, since doing so will be essential if it is to be able to seize the opportunities I have described (Coal Industry Bill, April 1980: Hansard)

There are two points of particular interest in this statement. First, the future of the coal industry was seen as strictly conditional upon its competitiveness and efficiency. Even though this debate took place at a time of continuing crisis in world oil markets, there was no suggestion of an unconditional commitment to the coal industry, still less to any kind of output target. Second, the industry should, as soon as possible, stand on its own feet, independent of government assistance. In its first year of office, the new Conservative government therefore clearly signalled that the need for the NCB's commercial viability overrode the energy policy 'rhetoric' on the strategic importance of coal; and this was made abundantly clear to NCB management.

It might be argued that the new arrangements were merely an extension of the arrangements put in place by the previous Labour government. Following Coal Industry Tripartite discussions in 1977, a financial framework had been agreed under which the industry would have the objective of long-term competitiveness, with the NCB aiming to provide 50 per cent of capital expenditure over the following five to six years from self-generated funds, and the system of EFLs for nationalised industries was inaugurated. The difference with the new Conservative government was the explicit rigour with which these principles were followed. There was a further change of symbolic importance. Under the Labour government, the NCB had regarded grants as part of its income within an annually-negotiated 'contract' with government. Under such a system, it was possible to declare an overall surplus after grants, and, for this to be characterised as a 'profit'. The use of a 'deficit grant' under the new regime made it impossible to put such a favourable gloss on the NCB's finances.

There was another important element in the NCB's new financial regime. In October 1979, the Board entered into the Joint Under-standing with the Central Electricity Generating Board (CEGB), covering a five-year period, whereby the CEGB undertook to use their best endeavours to take all suitable NCB coal up to a total of 75m. tonnes a year, and would adjust their coal import plans to enable them to do so, provided that NCB pithead prices did not rise at a rate greater than general inflation. As these arrangements were between nationalised industries, and affected their prices and financial results, the at least tacit agreement of government was required. But the initiative for the arrangement came jointly from the NCB and CEGB, both of which were seeking forward stability at a time of upheaval in world energy markets. This was in spite of considerable opposition from the Treasury, which thought that the large increase in oil prices occurring at that time provided an opportunity for the NCB to improve its finances by further coal price increase (Private communication: M J Edwards). The effective capping of the prices of some two-thirds of the

NCB's output (the remainder being more subject to direct competition with other fuels) provided an essential component of financial discipline: the NCB was required to improve its financial results, but now would have to do so by reducing costs rather than increasing prices.

However, no sooner had the NCB's new financial framework been put in place than it began to run into problems, due largely to the onset of economic recession and its effects on energy demand. In its Annual Report for 1979/80, published in July 1980, the NCB had pointed to the forthcoming difficulties involved in meeting the new financial objectives 'which were negotiated before the onset of the recession. The coal industry is too large and too complex to be able to change rapidly, and in responding to the challenge, the Board hope the government will provide them with a measure of short-term flexibility, particularly in regard to the treatment of coal stocks, which can have a disproportionate effect on the cash situation' (Chairman's Statement). While in retrospect this might seem like special pleading, the problems became real enough. Between 1979/80 and 1980/1, total UK primary energy demand fell by 25m. tonnes of coal equivalent (m.t.c.e), the total fuel required for electricity generation by 7 m.t.c.e., and total UK coal consumption by 8m. tonnes. This seriously disturbed

Table 1.3: UK Coal Supply and Demand 1979/80 and 1980/81. Million Tonnes

	1979/80	1980/1
Supply		
NCB output	123.3	126.6
Non-vested output	1.6	1.8
Imports	5.1	7.3
	130.0	135.7
Demand		
UK coal consumption	128.4	120.3
NCB Exports	2.5	4.8
	130.9	125.1
NCB Sales	125.1	117.7
Stock Change		
NCB stocks	-2.0	+8.8
Consumers stocks + balance	+1.1	+1.8
Total	-0.9	+10.6

Source: *NCB Annual Reports*

the balance between supply and demand for NCB coal, particularly as output and imports were on a rising trend. (See Table 1.3). Thus, in spite of an increase in exports, the NCB overproduced by nearly 9m. tonnes in 1980/1; and the increase in stocks had a significantly adverse effect on the NCB's External Financing Requirement in that year.

In 'real' terms, the EFR in 1980/1 represented an increase of 10 per cent over 1979/80, compared with the 7 per cent reduction envisaged in the new financial framework. Against this background, the task of reducing external finance, and achieving an overall break-even with only 'social cost' grants was looking increasingly difficult to achieve. After the NCB's Medium Term Development Plan (MTDP) for 1980/1 to 1984/5 was sent to the Secretary of State in March 1980, it became clear to the Board that the Plan's projections would be falsified by the recession and its effects on coal demand. The NCB instituted a 'Coal Policy Review', which concluded that, provided coal imports could be phased out, coal sales of 120m. tonnes (including 5m. tonnes of exports) could be maintained over the following five years. The projected financial outcome, however, was worse than the assessment made at the time the new financial framework was agreed in August 1979. The Board therefore examined the options that would be necessary if it was to recover its prospects of reaching its financial objectives. It concluded that an accelerated programme of closures of uneconomic, short-life collieries would be required. (Monopolies and Mergers Commission: A Report on the Efficiency and Costs in the Development, Production and Supply of Coal by the NCB, Cmnd 8920: June 1983, paras. 5.10 and 5.11)

Colliery Closures: The Retreat of February 1981

Throughout the autumn of 1980, Derek Ezra, the Board Chairman, had been warning David Howell that with the fall in coal sales, unless more public funding was available, pit closures were unavoidable (Lawson pp.140–1); and on 28 January 1981, Ezra briefed the Prime Minister on the closure plans. As Margaret Thatcher later wrote 'I agreed with him that with coal stocks piling up and the recession continuing there was no alternative to speeding up the closure of uneconomic pits' (Thatcher p.140).

The Board's analysis of the optimal way in which the industry's financial problems should be resolved, and the government's acceptance of that analysis, meant that there was now an industrial relations crisis in the making. Arthur Scargill, the hero of the militants within the

NUM, and leader of the Yorkshire miners, was anxious both to confront Mrs Thatcher's government and to ensure that he was successful in the campaign to succeed the moderate Joe Gormley as national President of the NUM in an election due before the end of 1981. Scargill held a ballot in Yorkshire in January 1981 on the principle of fighting pit closures, and secured an 86 per cent 'Yes' vote (Paul Routledge: *Scargill*, p.104). The NCB management were being driven into a corner. The government's attitude appeared to be: 'Stick to your financial objectives, but we don't want any trouble with the NUM.' It was therefore with a mixture of exasperation and trepidation that the Board sought a meeting with the unions on 10 February 1981, and put forward a 'four-point plan' designed to maximise sales by keeping prices competitive; increase efficiency and reduce unit costs; maintain a high level of investment in new and replacement capacity; and 'bring supply and demand into better balance by maximising sales, expanding output at pits with viable reserves, and diminishing capacity where realistic reserves were exhausted or where, for geological or other reasons, there could be no long-term financial contribution from a pit' (Ashworth p.416). The last point – expressed as a byzantine euphemism – was the heart of the matter. The Board had hoped (with the acquiescence of Joe Gormley) to avoid mentioning the number of closures, and to have a decentralised implementation, with Area Directors discussing the implications at local level. But in the clamour, this line was not held. Without warning, Derek Ezra told the unions that about 10m. tonnes of high cost capacity would have to be closed as soon as possible. This caused uproar. The National Executive of the NUM voted unanimously for a strike ballot, and as the NCB gave out details of which pits were to close (initially 23 collieries) unofficial strikes broke out which at the time looked likely to escalate into a full-blown official strike. Arthur Scargill said, 'Mrs Thatcher has been out to get the miners since 1972 and 1974. If she throws down the gauntlet, I can assure her of one thing: we will pick it up' (Routledge p.105).

The NUM demanded the recall of the Tripartite machinery, which had been set up by the Labour government in 1974 to launch 'Plan for Coal'. Initially a meeting was arranged for 23 February, but with the escalation of unofficial action in the coalfields, this was advanced to 18 February. Mrs Thatcher decided power station coal stocks were insufficient to withstand an all-out strike: 'There is no point in embarking on a battle unless you are reasonably confident you can win. Defeat in a coal strike would have been disastrous.' On the morning of 18 February, she met with David Howell 'to agree on the concessions which would have to be offered to stave off a strike' (Thatcher p.141); and on the evening of 18 February, at a meeting with the Mining Unions and the NCB management, Howell explained what the government had in mind, but in such terms that Gormley was able to announce to the media the government's 'total capitulation'. On the following day, David Howell made a statement in the House of Commons in which he reported that at the 18 February meeting

> three main points were raised – closures, financial constraints, and coal imports. I said that the Government were prepared to discuss the financial constraints with an open mind and also with a view to movement. The chairman of the National Coal Board said that in the light of this the Board would withdraw its closure proposals and re-examine the position in consultation with the unions. I accordingly invited the industry to come forward with new proposals consistent with 'Plan for Coal'. As regards imports, I pointed out that these would, in any case, fall this year from their 1980 levels. The industry representatives said that they wished to see this figure brought down to its irreducible minimum. I said that the Government would be prepared to look, with a view to movement, at what could be done to go in this direction (Hansard 19 Feb. 1981).

Given this statement, the NUM managed to get the local strikes stopped. In the last week in February, at meetings between the government, NCB and unions, the NUM pressed for large increases in government funding. The financial framework embodied in the Coal Industry Act 1980 was effectively abandoned. The government put pressure on the CEGB virtually to phase out imports.

The resulting increase in government funding was very large. The out-turn in 1981/2 as compared with 1980/1 is shown in Table 1.4.

The events of 'February 1981' had been very damaging for the government, particularly in terms of its standing within the Conservative party. As Nigel Lawson was later to write: 'The February

---

Table 1.4: NCB External Financing Requirements: 1980/1 and 1981/2. £million.

	1980/1	1981/2
Government grants	254	575
Additional loans	586	902
Total external finance required	840	1477

Source: MMC Cmnd 8920: Appendix 3.16

(1981) setback had forestalled any immediate hope of stemming the (NCB's) financial haemorrhage. Instead of trading new investment for old pit closures, as planned by David Howell, the government was now committed to subsidising old pits as well as investing in new ones' (Lawson p.144). He also wrote at the time to Geoffrey Howe that the events of February 1981 had shown beyond any reasonable doubt that the government would make no progress towards its aim of creating a commercial coal industry 'until we deal with the problems of monopoly union power' (Lawson p.142). Mrs Thatcher also had no doubt about the high political cost involved in avoiding an NUM strike in the winter of 1981. Yet while this episode increased the government's caution in its dealings with the coal industry, it also served to increase Mrs Thatcher's determination in due course to break the power of the NUM. As she later said in her memoirs: 'We would have to rely on a judicious mixture of flexibility and bluff until the Government was in a position to face down the challenge posed to the economy, and indeed potentially to the rule of law, by the combined force of monopoly and union power in the coal industry' (Thatcher p.143). The events of February 1981 also greatly increased the antipathy of the government to the NCB's senior management. There were reports of suspicion in government circles that in fact the whole crisis had been engineered by the NCB, in collusion with the 'moderate' faction of the NUM, to extract further financial concessions from the government to avoid the rundown of the industry. But what happened was due less to conspiracy than to miscalculation at a time when the Board had been caught between the rigour of the new financial framework introduced by government in 1979/80, and the increasingly militant opposition of the NUM to the closure of surplus uneconomic capacity.

However, the generous financial settlement of 1981 proved to be of no lasting value. The industry's fundamental problems remained un-resolved. 'February 1981' by critically strengthening the government's resolve to seek any opportunity to achieve its fundamental aims for radical change in the coal industry, can be seen, at least in retrospect, as one of the defining moments in the deep-mined coal industry's subsequent decline and fall.

CHAPTER 2
THE GREAT STRIKE

A Hiatus in Coal Policy

The NCB's reaction to the events of February 1981 gave the strong impression that it welcomed the government's capitulation, and the apparent closing off of the policy option of large-scale colliery closures, although this was by no means a unanimous sentiment among the senior management of the industry.

In his Chairman's statement in the NCB's Annual Report for 1980/1 (published in July 1981), Derek Ezra stated that 'any financial framework must be compatible with the increasing use of coal, improving output and productivity at long-life collieries, and continuing investment in new mining capacity beyond 1990'. The NCB endeavoured to reflect these principles in a new 'Development Plan to 1990', submitted to the Secretary of State in July 1981. This aimed to show that, given some recovery in demand, the NCB would be able to cover all its revenue costs, including interest charges, other than 'social' costs by 1987/8, and fund half its investment programme from internal sources. In physical terms, the main outlines of the Plan reflected the NCB's view that, following the events of February 1981, the principles of 'Plan for Coal' (interpreted in ways acceptable to the Mining Unions) should be seen to continue to apply, with investment continued at £730m. per annum, and with substantial additional output being produced by new investment, as well as the completion of existing schemes (notably the Selby coalfield), but with only a very gradual reduction (of 2m. tonnes per annum) in the output of 'short-life' mines, such that there would be a small net increase in total output over the decade. But this once again raised the question of how supply and demand were to be balanced, and some inroads made into excessive stocks. Here the Plan provided for substantial increases in supplies to UK power stations well above those provided in the 'Joint Understanding' with the CEGB, but also projections of a substantial increase in exports, which were effectively being used as the balancing item (See Table 2.1).

These physical projections were associated in the Plan with a huge improvement in financial results over the decade, moving from a

Table 2.1: 'Development Plan to 1990' (July 1981). Projections of Output and Sales. Million Tonnes

	1981/2	1982/3	1983/4	1984/5	1985/6	1986/7	1987/8	1988/9	1989/90	1990/1
Output										
(a) Approved investment only										
New mines	1	1	3	4	7	9	11	12	12	13
Long life mines	88	90	91	91	90	89	89	87	85	84
Short life mines	20	18	16	14	12	10	8	6	4	2
	109	109	110	109	109	108	108	105	101	99
(b) New investment	-	-	-	1	3	4	5	8	13	15
Total deep mines	109	109	110	110	112	112	113	113	114	114
Opencast etc	15	15	16	16	16	16	16	16	16	16
Total	124	124	126	126	128	128	129	129	130	130
Sales										
Inland	112	113	114	115	115	117	118	120	122	124
Export	9	9	11	14	15	13	13	11	10	8
	121	122	125	129	130	130	131	131	132	132
Stock Change	+3	+2	+1	-3	-2	-2	-2	-2	-2	-2

Source: MMC (Cmnd 8920) Tables 5.5 and 5.3

projected net loss before deficit grants of £457m. in 1981/2, to a projected net profit of £680m. in 1990/1 (See Table 2.2). This enormous turn-round in projected financial results depended crucially on the deep-mined results. The NCB assumed that labour productivity would increase by 3 per cent per annum and that (at 1981/2 money values) average colliery operating costs would be reduced from £38.59 in 1981/2 to £34.93 by 1990/1 (a real reduction of only 1 per cent per annum) (MMC Cmnd 8920 Table 5.1). The main source of projected improvement was the assumption that, once the restraints of the 'Joint Understanding' came to an end in March 1985, average prices would rise by some 3 per cent per annum in real terms, reflecting the view, common in official circles at the time, that international steam coal prices would be on a rising trend as world demand increased in order to reduce dependence on oil: a view which subsequently came to be totally discredited.

The 'Development Plan to 1990' was a thoroughly unsatisfactory document. It had been drawn up to be consistent with a 'minimalist' approach to the closure of high-cost capacity and manpower reductions. Under these circumstances, it was almost impossible to project a satisfactory financial result, even over a period of years, without making over-optimistic assumptions about power station demand and coal prices from 1985, and postulating large (and uneconomic) export sales.

The Department of Energy was very critical of the Plan, but did not suggest any sensible alternative way forward. At a meeting with senior officials, the author asked whether the Department would prefer the NCB to raise again the closure of uneconomic capacity. The question was met with silence. The appeasement settlement following February 1981 led to a hiatus in government policy on the industry's fundamental problems.

This did not mean that there had been any second thoughts by the government on the political objectives for coal, particularly as Nigel Lawson succeeded David Howell as Secretary of State for Energy in September 1981, and proceeded to articulate a distinctively 'Thatcherite' rationale for policies towards the energy industries generally. He was 'determined to break the "dirigiste" mentality that pervaded both the Department of Energy and the nationalised energy industries' (Lawson p.163). In 1982 he set out his thinking more fully in a speech at a conference of the International Association of Energy Economists at Churchill College, Cambridge, in which he derided policy-making by means of long-term forecasts and central planning which allocated future demand among primary fuels. Rather, he

Table 2.2: 'Development Plan to 1990' (July 1981). Projections of Revenue Results. £ million 1981/2 money values

	1981/2	1982/3	1983/4	1984/5	1985/6	1986/7	1987/8	1988/9	1989/90	1990/1
Operating Profit (loss)										
Collieries	(359)	(365)	(299)	(243)	(93)	60	220	356	511	654
Opencast	169	164	178	182	184	201	212	228	242	259
Other activities	(12)	5	9	11	12	10	8	7	6	4
	(202)	(196)	(112)	(50)	103	271	440	591	759	917
Social cost grants	106	118	111	105	101	91	91	90	89	87
Interest	(361)	(362)	(389)	(390)	(424)	(452)	(421)	(451)	(408)	(324)
Additional grants to break-even	457	440	390	335	220	90	-	-	-	-
Net surplus	-	-	-	-	-	-	110	230	440	680

Source: MMC (Cmnd 8920) Table 5.1

argued, energy should be treated as 'a traded commodity' (Lawson pp.164–5).

The MMC Reference: Analysis without Solutions

How this market-related thinking should, in precise terms, impinge on the activities of the NCB remained unclear for some time, notwithstanding the arrival of Nigel Lawson. Rather than give clear guidance to the NCB, the government decided to refer the NCB's affairs to the Monopolies and Mergers Commission (MMC). The formal terms of reference were published on 3 March 1982, requiring the Commission to

> investigate and report on the question whether the Board could improve its efficiency and thereby reduce its costs, with particular reference to:
> (a) the extent, if any, to which its operating costs can be contained or reduced;
> (b) its system of internal cost control;
> (c) its purchasing policies;
> (d) its methods of controlling its stocks of stores and materials;
> (e) the planning and appraisal of new investment;
> (f) the management, supervision and control of investment projects.
>   (MMC Cmnd 8920 Para. 1.1)

This investigation would continue until December 1982.

The terms of reference were carefully chosen to avoid any explicit focus on the problems of uneconomic capacity and over-manning, and gave the impression that the industry's problems could be solved by better management procedures. The management issues raised were indeed important, but not germane to the central issue, which was surplus production of uneconomic output. The NCB team dealing directly with the MMC sought to give this central issue the greatest possible prominence in the analysis of material presented, much of which was subsequently incorporated in the MMC's report (Cmnd 8920). In 1981/2, the NCB was overproducing by at least 10m. tonnes relative to inland demand (that is, disregarding uneconomic exports). But in that year the most expensive 10 per cent of deep-mined output, at collieries with production of 10.8m. tonnes, incurred operating losses of £263m. (£24.3 per tonne), which more than accounted for the aggregate operating loss for all deep mines of £226m. (excluding operating grants) (MMC Cmnd 8920 Table 8.4). Over the six years to 1981/2, the results of the collieries operating throughout that period are shown in Table 2.3.

Table 2.3: Classification of Continuing Collieries by Profitability. 1976/7–1981/2

	Number of Collieries
Consistently profitable	23
Consistently unprofitable	59
'Variable' results	116
	198

Source: Derived from MMC Cmnd 8920 Table 8.5

Of the consistently unprofitable collieries, 39 had losses of over £10/tonne (around 30 per cent of sales prices) in 1981/2. And collieries with losses of over £15/tonne employed 40,000 men. It was the manpower dimension to surplus uneconomic output which was to provide the greatest difficulty both for the NCB management and for the government. On the other side, there were 36 collieries which made operating profits in at least five out of the six years to 1981/2, and in 1981/2 produced 32.1m. tonnes, or only 30 per cent of total deep-mined output; and only about a fifth of the industry's 218,000 mineworkers were employed at consistently profitable pits. (NCB data in MMC Cmnd 8920 Vol II).

Short of closing more high-cost capacity, the NCB had to consider a number of other 'short-term regulators' to achieve a better balance between supply and demand. (Dealt with in MMC Cmnd 8920 paras. 5.56–5.68.)

Avoidance of imports. But, following the settlement of 'February 1981' these were already close to minimum levels. Indeed, the CEGB told the MMC that, if they were free agents, they would seek to increase imports of steam coal which, given the continuing high price of fuel oil, represented the main alternative to NCB coal. International seaborne trade in steam coal (mainly for power stations) had received considerable impetus from the two oil price 'shocks', and doubled between 1978 and 1984. Although only representing a small proportion of total steam coal production worldwide, this trade was becoming of growing importance in Western Europe and the Pacific Rim, not least by acting as a point of reference for 'market prices'. However, in the early 1980s, UK import facilities for power station coal were limited to about 12m. tonnes per annum, with most of this in small vessels 'trans-shipping' coal delivered in bulk carriers to Rotterdam, and there were no port facilities dedicated to the direct import of power station coal in large vessels. This, together with the cost of rail transport within the UK, gave the NCB a significant geographic advantage in terms of delivered prices at the main coal-fired power stations in the Midlands and Yorkshire. Evidence by the CEGB to the MMC in October 1982 gave a comparison of delivered prices for NCB coal and imported coal (both 'contract' and 'spot' prices) at four selected power stations in a range of locations (See Table 2.4). This data showed that NCB prices were competitive against imports sold under 'contract', except on the Thames Estuary (that is those power stations most distant from the coalfields). Although the comparisons were less favourable if 'spot' import prices were used, the NCB pointed out that the spot prices quoted by the CEGB represented very limited tonnages in world trade, and any attempt to import large tonnages would increase import prices (particularly spot prices). Nevertheless, while NCB coal was generally competitively priced on the central coalfields (where most of the coal was consumed), the adverse position at power stations away from the coalfields represented a threat to the total volume of NCB sales. For this reason, in November 1982 the 'Joint Understanding' with the CEGB was modified so that tonnage supplied over 70m. tonnes would attract a discount to align with import prices (*NCB Annual Report 1982/3*). Additional discounting had also been introduced in the industrial market to safeguard sales against coal imports or other fuels.

Table 2.4: Delivered Price of Power Station Coal 1982. (1982 pence per GJ)

	Thames Estuary	Didcot (Berks)	Ratcliffe (Notts)	Fiddler's Ferry (Cheshire)
Typical NCB coal	190	190	173	175
Imported coal				
Contract	175	200	196	198
Spot	141	165	162	163

Source: MMC (Cmnd 8920) Table 4.11

Additional exports. In 1981/2 the NCB exported 9.4m. tonnes, compared with only 2.5m. tonnes in 1979/80, and, as we have seen, contemplated further increases to 15m. tonnes by 1985/6 in the July '81 Development Plan. In economic terms, the NCB sought to justify such an approach in that, although the pithead realisation for exports was only about £25 per tonne (compared with £37 for sales to the CEGB), this price was higher than the short-run avoidable costs at deep mines

(£15) and the average cost of opencast sites (£20), and that therefore exporting was 'profitable'. However, this argument only held if failure to export would in fact cause output to be curtailed at continuing collieries or at opencast sites. In more conventional economic analysis, the industry's marginal costs would be those avoided by the closure of the highest-cost collieries – well over £50 per tonne. If the lower revenue per tonne available from exports had been directly attributed to the highest-cost pits, their losses would have been even greater. Exports made sense only because the closure of more high-cost capacity was not a practicable option at that time.

Reduction in the scale of colliery operations. The NCB generally ruled out reduction in output at continuing collieries, for example by reducing the number of coal faces, as much of the cost of operating a colliery did not vary with output, with a result that avoidable marginal costs were less than half of average operating costs, and less than the revenue obtainable in the marginal market, namely exports.

Reduction in opencast production. Opencast output had a much lower average cost than deep mines, although there was limited scope to postpone production if incremental (i.e. avoidable) output was valued at export prices (which had become NCB standard practice). On the other hand, much opencast output was either of special coals (particularly anthracite), which did not compete with surplus power station coal, or was used for blending to make lower-quality deep-mined coal more acceptable to power stations (particularly in respect of chlorine content).

Rephasing of investment. Reducing capital expenditure was largely inappropriate as a means of correcting short-term excess supply. Much of the NCB's capital expenditure was absorbed by large long-term projects, often subject to contract commitments, and by non-output related projects to reduce costs. Reductions in 'operational' capital (for example coal face machinery) would curtail output at continuing collieries, with the adverse effects on average costs already noted.

As the industrial and economic case for a radical reduction in high-cost capacity became clearer, so the organisation of militant NUM opposition to such an approach gathered force. In March 1982, Arthur Scargill assumed the Presidency of the NUM in succession to Joe Gormley. There was no doubt that he approached his task in terms of the politics of class warfare. In the national strike of 1972 he had been the hero of the militant Left in the battle of the Saltley coke depot where 'flying pickets' had outfaced the local police. Scargill was to say of this use of pickets: 'we took the view that we were in a class war ... we had to declare war on them and the only way to declare war was to attack the vulnerable points ... we wished to paralyse the nation's economy' (Quoted in Hattersley: *Fifty Years On*, p.219). On becoming President, Scargill saw his task as creating the conditions for an all-out strike against the government. At the NUM 1982 annual conference, a 31 per cent pay claim was adopted, and constitutional changes made which removed the previous 'moderate' majority in the NUM National Executive. In October 1982, a special delegate conference rejected the NCB's 8.5 per cent pay offer, and imposed an overtime ban as a prelude to a strike ballot on 28 October. In the meantime Scargill addressed a series of rallies in which he sought to link the rejection of the wage claim with the issue of colliery closures on the grounds that the Board was 'looking for the slightest sign of weakness so that the mad dogs of pit butchery can be unleashed'. But the rank and file rejected Scargill's strategy by 61 to 39 per cent. (Routledge pp.118–120).

Incensed by this ballot result, Scargill made public allegations against the NCB that they had hidden 'hit-lists' of collieries to be closed. This matter was taken up by the House of Commons Energy Select Committee (Report 'Pit Closures' HC 135 Dec. 1982). In his opening oral statement to the Committee, Scargill argued that leaked NCB internal documents 'indict the National Coal Board and brand them guilty of duplicity and lying to the NUM and the public at large', and that the Committee should recommend the dismissal of the Board (para 8). Scargill claimed that 'over an 8 year period the Board appears to be planning to close 95 pits with the loss of 70–100,000 jobs', these figures being derived from confidential internal papers prepared by the NCB's Statistics Department analysing the results of 'short-life' and 'unprofitable collieries' (para. 5). These allegations, against the background of 'February 1981', caused the NCB some anxiety, but they claimed (which was no less than the truth) that these analyses of short-life, uneconomic pits do not constitute a 'hit list' in the sense of a list of named pits about whose closure firm decisions had been taken. The Committee accepted this interpretation (para 10), and went on strongly to reject Scargill's assertions that coal should be mined to physical exhaustion, whatever the financial loss involved (para 21), that the Mining Unions should have a veto over closure decisions (para 22) and that UK coal output should be built up to 200m. tonnes by 2000 (para 35). Moreover, the Committee endorsed what the NCB had sought to emphasise to the MMC about the 'uneconomic tail': 'There can be no doubt that the very high cost of this and, at least in the short term, surplus capacity unfavourably distorts the coal industry's financial position, and imposes a considerable drain on public funds' (para 30).

The Committee's report was, however, less helpful on what should be done about surplus, uneconomic capacity. The Committee's comments that they had been 'puzzled by the Board's excessive defensiveness about the status of the leaked documents' (para 11) and that more open disclosure in future of the Board's internal analyses would be helpful (para 13), seemed politically naive in the circumstances of the time, when it was implicit in the Board's terms of reference that they should do nothing to provoke a national strike (as a list of intended closures had almost done in February 1981). The Committee went on to say on the one hand that 'the Board must take steps to bring its capacity more into line with existing and expected demands for coal' (para 38); while on the other hand that the Department of Energy 'must take a more active and less purely responsive role in ensuring, firstly, that the Board has full Government support in any effort it might make to reduce the industry's surplus capacity (especially that with the highest costs), and secondly, that a new strategic plan is developed for the coal industry at the earliest opportunity', and that this 'new corporate plan should be published and openly discussed' (para 39). There was never any serious chance of such advice being taken at that time.

In December 1982, the government was not prepared to risk a national strike. In that month, it received the MMC's Report on the NCB, the contents of which might have been considered inflammatory. In the event, publication of the Report (Cmnd 8920) was delayed until after the government's election victory in June 1983. However, the Report's findings were not as controversial as they might have been. In the first place, only one of the 48 formal recommendations concerned the central question of the problem of surplus uneconomic output (Recommendation 4 said that, over the longer term the NCB should not rely on low-price exports to reduce surpluses of production). The other recommendations concerned internal NCB procedures for budgeting, planning, capital investment appraisals and control, operating cost control, purchasing and stores and opencast management. Secondly, although the MMC analysis of the central problem of the industry closely followed that of the NCB, the Commission were short on clear recommendations as to what specifically should be done. 'We have noted ... the substantial proportion of output produced at heavily loss-making collieries. The most obvious way for the NCB to reduce excess supply would be to close the pits concerned. Although ... the NCB has been heavily constrained as regards closure of uneconomic capacity' (para. 5.64). Thus, 'When we questioned the NCB as to why it had failed to achieve the level of closures envisaged in the Tripartite Report of 1974 and why it had assumed only a modest level of closures in its current medium-term development plan, we were told in both cases that the necessity of maintaining good industrial relations precluded faster progress' (para. 8.25). 'The NCB is facing two crucially important problems, separate but closely related. They concern over-capacity and high cost pits' (para.19.13) 'We estimate that if capacity could be reduced by 10 per cent and the reduction concentrated on these pits with the largest operating losses per tonne, the NCB's finances would be improved to the tune of £300m. per annum' (para.19.16) 'The longer the problems are left the worse they will become. Unless there is a significant reduction in the number of high cost pits, the NCB's finances will deteriorate even further' (para.19.22) But, after this admirably clear analysis, the Commission concluded lamely that, 'It would not be appropriate for us to define precisely the way in which the NCB should reduce excess capacity in high cost collieries. We do not attempt to specify by precisely how much capacity should be reduced, in what period the reduction should be made, or what individual collieries should be involved. It is for the Board to face the problem and take the necessary action' (para. 19.23). The buck was passed back to the NCB.

The government was, however, happy to accept the MMC Report on this basis, which preserved the fiction of the arm's-length relationship between the government and the NCB, and which enabled the Report to be presented as an indictment of NCB management, rather than as a matter of government policy. For its part, the NCB noted wryly in its initial response to the Commission's Report in July 1983 that 'The Commission agreed with the Board's own view that surplus production from high-cost pits was the crucial problem preventing the industry from becoming viable. The Board stated that they were fully aware of the issues involved and had no illusions about the scale of the problems to be faced' (reported in *NCB Annual Report 1983/4*, p.9).

A Recipe for Conflict

What, therefore, was the state of the industry in 1983?

Coal retained its share of the UK primary market at around 36 per cent, as it had done since the first OPEC price shock in 1973/4. Yet

UK primary fuel demand fell by 44m. tonnes of coal equivalent between 1979 and 1982, and this was reflected in a sharp reduction of 18m. tonnes in the consumption of coal. (See Table 2.5)

Table 2.5: UK Coal Consumption: 1979/80 to 1982/3. Million Tonnes

	1979/80	1980/1	1981/2	1982/3
Power stations	89.1	87.7	85.3	80.8
Coke ovens	14.4	11.3	11.2	10.0
Domestic	10.3	8.5	8.6	8.0
Industry etc.	14.6	12.7	11.9	11.6
	128.4	120.3	117.0	110.4
Of which imports	5.1	7.3	4.2	3.4
NCB market share	96%	94%	96%	97%

Source: *NCB Annual Reports*

Over half of this reduction occurred in non-power station coal markets. Falling coking coal sales were the result both of the failure of the earlier BSC expansion plans to materialise, and also of the virtual absence in the UK of the very high quality coking coal now required by the steel industry in an increasingly competitive market. Although the NCB still entertained ambitions for an increase in industrial sales, coal's position was affected by falling industrial energy demand (often involving the scrapping of obsolete coal-burning plant) and by increasing competition from gas; and the domestic solid fuel market was in long-term decline under the impact of Clean Air Legislation and rising sales of more convenient natural gas.

But more serious for the NCB was the decline of consumption in power stations by over 8m. tonnes between 1979/80 and 1982/3. Increasingly, the UK coal industry depended upon sales to power stations, which represented about 70 per cent of total NCB sales. But coal (which was nearly all NCB coal) had also become the dominant fuel for power generation, representing over three-quarters of the total fuelling requirement of power stations. (See Table 2.6)

This symbiotic relationship between coal and power generation reflected the realities of the primary energy market at the time. Notwithstanding the steady but modest rate of increase in the contribution of nuclear power, coal was able to consolidate its position by progressively displacing fuel oil. (There was no suggestion that natural gas, then regarded as a scarce 'premium' fuel, should be used

Table 2.6:   Fuel used for Power Generation*: 1979/80 to 1982/3. Million Tonnes
Coal Equivalent

	1979/80	1980/1	1981/2	1982/3
Coal	88.8	87.3	85.0	80.3
Oil	13.1	8.7	8.6	5.7
Nuclear	12.1	12.0	12.2	16.2
Other	2.8	1.8	1.7	3.1
	116.8	109.8	107.5	105.3
Coal's share	76%	79%	79%	76%

*   CEGB and SSEB

Source: *NCB Annual Reports*

for power generation). Coal's competitive advantage had been made possible by the very large increases in fuel oil prices following the OPEC crude oil price increase of 1973/4 and 1979/80; so that, although coal prices also rose substantially in real terms during the 1970s, and remained at a high level thereafter, coal prices per GJ remained only about 60 per cent of oil prices – a differential more than sufficient to make coal the preferred fuel for electricity generation. Coal consumption at power stations had declined between 1979/80 and 1982/3 by over 8m. tonnes, not because coal was uncompetitive, but because the recession and decline in industrial activity had reduced electricity demand and hence the total fuel requirement for power generation.

Although between March 1980 and March 1983, the NCB reduced the number of operating collieries by 28, and the labour force by 30,000 men, reductions in output did not match the fall in demand, so that over that period the NCB was seriously overproducing relative to UK coal demand by some 13m. tonnes a year, that is, before taking account of (largely uneconomic) exports. (See Table 2.7)

Even though an additional 14m. tonnes of coal were exported over the period, total UK stocks of coal almost doubled between March 1980 and March 1983, by which time they were equivalent to nearly half of annual demand. (See Table 2.8)

Yet, looking beyond March 1983, this excess stocking position looked to be getting worse, given the continuing inhibitions on closures following the events of February 1981, even though imports were being restricted with government approval. Moreover, the problem of over-production was set to become even more severe as major projects already approved were completed. By 1982, over half of the 'incre-

Table 2.7:   UK Coal Consumption and NCB Output 1979/80 to 1982/3. Million Tonnes.

	1979/80	1980/1	1981/2	1982/3
UK coal consumption less imports	123.3	112.9	112.8	107.0
NCB production				
Deepmined	109.3	110.3	108.9	104.9
Opencast + Other	14.0	16.4	15.4	15.9
Total	123.3	126.7	124.3	120.8
NCB 'over production' *	-	13.8	11.5	13.8
(NCB exports)	(2.5)	(4.7)	(9.4)	(7.1)

*   NCB output, less UK coal consumption less imports (and before taking account of NCB exports)

Table 2.8:   Coal Stocks: 1979/80 to 1982/3. Million Tonnes

	1979/80	1980/1	1981/2	1982/3
Inland consumption	128.4	120.3	117.0	110.4
Year end stocks				
NCB stock	12.0	20.9	24.9	24.9
Consumers stock	15.8	17.5	18.6	28.3
Total stocks	27.7	38.4	43.5	53.3
Stock as % of consumption	22%	32%	37%	48%

Source: *NCB Annual Reports*

mental' low-cost output envisaged in 'Plan for Coal' to be available by the mid-1980s had still to be achieved. (See Table 2.9)

The largest project still uncompleted in 1982 was the Selby new mine complex. If the success of this major new investment was not to be prejudiced, the new output from Selby and other major projects already under construction would require further 'headroom' in the market of up to 25m. tonnes per annum, if over-production was not to increase dramatically. This unsustainable position could be resolved in a rational manner only by the accelerated closure of the uneconomic 'tail', which would give rise to strong opposition from the militant NUM leadership and probably a national strike.

Table 2.9: NCB Major Projects related to Capacity (Position in 1982). Million Tonnes 'Incremental Output' *

	Completed projects	Projects under construction
New mines	1.6	12.3
Existing mines	13.2	13.1
	14.8	25.4

Source: MMC (Cmnd 8920) Tables 3.11 and 3.12

* In the case of existing mines, 'incremental output' represents the difference between planned output on completion against output projected in the absence of major investment

Table 2.10: NCB Profit and (Loss) Account 1979/80–1982/3. £ million.

	1979/80	1980/1	1981/2	1982/3
Operating Profit/(Loss)				
Deep mines *	(122)	(107)	(226)	(312)
Opencast	110	157	157	192
Other mining activities	6	12	21	23
Total mining activities	(6)	62	(48)	(97)
Non-mining activities	26	17	5	3
Total Operating Profit/(Loss)	20	79	(43)	(94)
Social costs less grants	(17)	(29)	(61)	(49)
Sales of fixed assets	25	19	20	20
Trading Profit/(Loss)	28	69	(84)	(123)
Interest	(185)	(256)	(341)	(366)
Miscellaneous items	(2)	(20)	(3)	4
Deficit grant	159	149	428	374
Overall Surplus/(Deficit)	-	(58)	-	(111)

Source: *NCB Annual Reports*

* After crediting 'operating' grants

Crucial though the problem of over-production was, it was not the only one. The NCB's overall financial position continued to deteriorate after 'February 1981'. (See Table 2.10)

Deep-mine operating losses, which by 1982/3 equalled 8 per cent of turnover, were rising, and giving rise to an overall operating loss in spite of an increasing cross-subsidy from the operating profits of opencast and other mining activities. Moreover, the NCB's deep mines were not only losing large sums before any repayment of interest, but were at the same time still engaged in the very large capital programme associated with 'Plan for Coal'. In the two years 1981/2 and 1982/3, deep mines made virtually no internal contributions to this capital expenditure of some £700m. a year. (See Table 2.11)

Table 2.11: NCB Deep-mined Operating Profit and Capital Expenditure. 1981/2 and 1982/3. £ million.

	1981/2	1982/3
Deep-mined operating Profit/(Loss)	(226)	(312)
Plus Historic Cost Depreciation	246	294
Deep-mined operating Profit/(Loss) before depreciation	20	(18)
Capital expenditure on fixed assets	694	716

Source: *NCB Annual Reports*

Although opencast profits contributed about a quarter of the funds required for deep-mined investment, three-quarters was being provided by government borrowing; and because interest payments to government in 1981/2 and 1982/3 were more than matched by government deficit grants (see Table 2.10), this borrowing was effectively interest free. It was clear that, at some point in the future, a radical reconstruction of the NCB's finances would be required. For the present, the MMC noted (Cmnd 8920 para 3.133) that 'the NCB has been financed almost entirely by loan capital and government grants during the last ten years. The loan capital has come mainly from the National Loans Fund (NLF), and has usually been at fixed rates of interest for a relatively long period ... It might be easier to evaluate the financial performance of the industry in future if it had funding arrangements closer to those of a commercial organisation.' But this was largely beside the point. Even if the NCB had been financed wholly by some form of government equity, with no 'dividend' payable on these funds until some time into the future, the deep-mine operation would still have been heavily loss-making on revenue account, with negligible internal funding of investment.

Deep-mined operating results had been adversely affected by downward pressure on average revenue. This was a result of a number of factors. The share of high-priced markets, particularly for domestic coal, was falling, while the share of lower price power station coal was rising. While the price of most power station coal was linked to general inflation under the terms of the 'Joint Understanding' with the CEGB, in 1982/3 a lower-price tranche was introduced to combat lower import prices. In the industrial market, price rebates were widely used to protect business from competition from gas, oil and imported coal. The increase in exports, as part of the NCB's response to over-production in the home market, also had a materially adverse effect on average revenue, since export realisations were over 30 per cent below inland sales (MMC Cmnd 8920 para. 4.61).

Although average deep-mined operating costs were broadly stabilised in 'real terms', over the 1980/1 to 1982/3 period (as compared with the 'real' increase of nearly 50 per cent between 1972/3 and 1979/80), there was no progress in matching falls in average revenue – still less in making inroads into the operating losses. Some improvements were being made in labour productivity, particularly at the coalface, but overall the result was modest. (See Table 2.12)

Table 2.12: Labour Productivity. NCB Deep Mines. 1979/80 to 1982/3

	1979/80	*1980/1*	*1981/2*	*1982/3*	*% increase 1982/3 c.f. 1979/80*
Output per manshift (tonnes)					
Production/coalface	8.88	9.09	9.56	10.10	14%
Overall	2.31	2.32	2.40	2.44	6%
Output per man year (tonnes)					
Overall	470	479	497	504	7%

Source: *NCB Annual Reports*

Whether measured in terms of output per manshift or as output per manyear, labour productivity was growing at some 2 per cent per annum. But this was insufficient to offset rises in costs.

In spite of the high level of operating losses on deep-mined, and the modest growth in overall productivity, miners' wages remained high. In 1982, the NCB told the MMC that one of its guiding principles over a number of years had been to keep mineworkers 'high in the earnings league' (MMC Cmnd 8920 para.13.23). In 1982, average weekly earnings in coal mining were some 27 per cent above those in manufacturing, and miners were top of the 'earnings league', apart from the relatively small number of workers at oil refineries (MMC Cmnd 8920 Table 13.7). The NCB also told the MMC that 'part of its policy regarding the position of mineworkers' earnings related to those elsewhere is in recognition of the arduous nature of the miners' task and in order to ensure that the industry can recruit and retain the labour force required' (MMC Cmnd 8920 para.13.28). The MMC response to this particular argument was diplomatic:

> We have noted that in spite of increasing operating deficits and mounting stock piles of unsold coal, and against the background of reduced recruitment needs and a decline in natural manpower wastage, mineworkers' earnings have continued to increase relative to those in most other industries. We have no doubt that this is a reflection of the collective bargaining pressures in the industry. Nevertheless we are not satisfied that the industry's financial and economic positions have been accurately reflected in the (wage) agreements reached in recent years (MMC Cmnd 8920 para.13.50).

While fair up to a point, this verdict did less than justice to the great difficulties facing NCB management in settling annual wage claims without provoking strike action.

In March 1983, at a time of continuing overproduction and worsening of an already poor financial position, the government published objectives for the NCB which it had settled with the Chairman (Norman Siddall) in the following terms:

1. Although coal is one of the United Kingdom's major natural resources, in the Government's view the justification for coal production, like that for any other business, lies in the ability of those engaged in it to earn a satisfactory return on capital while competing in the market place. The basic objective for the National Coal Board therefore must be to earn a satisfactory return on its assets in real terms after payment of social grants. This return will be quantified in due course.
2. The NCB should aim at that share of the market which they can profitably sustain in competition with other fuels. The Board should not plan on any continuing tranche of sales which will not be profitable. The Board should bring its productive capacity into line with its continuing share of the market.
3. The Board's objective should be to ensure that over the next five years its operating costs (including depreciation and capital charges but excluding interest) per tonne of coal produced are reduced in real terms for deep-mined and for opencast production separately. (In this calculation the GDP deflator should be used). This objective will be quantified in due course ...

(NCB Annual Report 1982/3, p.3)

There were some points of obscurity here. While wishing to avoid cross-subsidy between markets, the statement carefully avoided saying that the NCB should not plan to have any continuing tranche of output produced at a loss. Nevertheless, it is clear that the government wished to emphasise at that time that the NCB must renew its efforts to eliminate surplus production and substantially improve its financial results.

Such efforts would, in all probability, have brought the NCB into collision course with the militant leadership of the NUM, which was determined to bring about a national strike. Yet as long as the NCB believed that it was central to its remit with government that it should not provoke such a strike, the power of the NUM would remain perhaps the crucial influence in the way the industry was managed. The NUM's power, which derived from the fact that the electricity industry was 75 per cent dependent on NCB coal, manifested itself in a number of ways: the high level of wages, the strong inhibition against the rundown of manpower by closing high-cost pits and acting as a general constraint on productivity by 'de-manning' at continuing collieries; and pervasively at operational level, local management's pre-occupations were as often centred on industrial relations issues as upon the technical and financial aspects of deep mining.

By 1983 there were intolerable tensions at the centre of the NCB's affairs: between pressures from the government to improve the industry's poor economic and financial performance, and continuing efforts to keep the peace with the NUM, whose leadership had adopted an impossibilist position that economic considerations should be absent from the coal industry's management, and was determined to mount a national coal strike against the government.

These tensions were to be resolved by the Great Strike.

How the Great Strike Began

In March 1983, Scargill failed for the third time to secure a ballot majority for a national strike, losing by exactly the same margin as in October 1982 (61 per cent to 39 per cent) (Routledge p.121). The NCB, under the wise and subtle leadership of Norman Siddall, who had been Chairman since July 1982, continued its policy of rationalisation 'by stealth', which it had followed since February 1981. Under this policy, most of the 23 collieries, whose announced closure had been the occasion of the government's humiliation in February 1981, were closed within two years and average manpower was reduced by 22,000. The main features of the policy were firstly, at all costs to avoid the publication of a nationally-determined list of closures, but to decentralise procedures to local NCB Area level, bringing forward proposals one-by-one for local agreement under the colliery review procedure; secondly, to follow the principle of 'no compulsory redundancies', and to encourage miners to opt for redundancy by offering generous severance terms; and, finally, to keep miners 'at the top of the earnings league', with substantial productivity-related bonus payments from incentive schemes. This approach made it much more difficult for the NUM at local level to mount effective opposition to colliery closures.

Moreover, the real (albeit insufficient) progress made during the brief Siddall Chairmanship, and the continuing inability of Scargill to win a clear-cut mandate for strike action, owed a great deal to Siddall's personal credentials with the work force. When he came to hand over the Chairmanship of the NCB to Ian MacGregor in September 1983, Norman Siddall urged the new Chairman to continue the previous policy of rationalisation 'by stealth', and had advised 'against the Cabinet picking a fight with the miners just for a show of strength' (quoted in Routledge p.128). This raises the question whether the continuation of the Siddall policy (but without Siddall) was a sustainable option, in view of the seriousness of over-production, or whether a final confrontation with Scargill was now inevitable, particularly as MacGregor's appointment was widely seen as provocative, given the way in which he had shed labour at BSC, and with his history of opposition to unions in the American coal industry.

There is plenty of evidence that Scargill was determined to mount a strike, even if the requirement for a majority in a national ballot of the NUM membership had to be circumvented. As early as March 1983, and before the national ballot was lost, he had suggested in a BBC radio interview that such a ballot was not required before the union could mobilise national support for a strike. On this view, individual NUM Areas could use Rule 41 to call strikes in their own coalfields, either through a local ballot or any other means sanctioned by their own (rather than national) rules, and such strikes could receive national support. If all Areas proceeded in this way 'this could amount to a national strike called by different, though still constitutional means' (Routledge p.140).

However, this was not NUM policy from the outset. Moreover, the closures issue became entangled with the annual wage claim. The Annual Conference in Perth in July 1983 passed resolutions both on the forthcoming wage claim, and on the union's opposition to pit closures and manpower reductions. Both resolutions called for ballots of the membership if satisfactory responses were not obtained from the NCB.

In September 1983, the NUM presented to the NCB a claim for a substantial wage increase, payment on a salary basis (thereby removing the performance related Incentive Scheme) and a reduction in working hours. Although government policy at the time was that public sector wage settlements should not exceed 4 per cent, the Board responded, after consultation with government, with a 5.2 per cent offer, which would maintain mineworkers' position at the top of the 'earnings league' (Ned Smith; *The 1984 Miners' Strike: The Actual Account*, p. 19). An NUM Special Delegate Conference on 21 October rejected the Board's wage offer as 'totally unsatisfactory', and reaffirmed its total opposition to pit closures and reductions in manpower. Although the NCB sought to establish that their wage offer was not conditional on acceptance of any level of closures, the NUM leaders continued to assert a connection between wages and colliery closures, and on this basis imposed a national overtime ban from 31 October.

In the meantime, MacGregor had been making his plans. In December 1983 he indicated to the government that he hoped to cut the workforce by 44,000 over the following two years, with 20,000 in 1984/5. Yet this was still to be carried out within the framework of Siddall's policy of closure 'by stealth', and without major industrial conflict. Decisions on closures would still be on a pit-by-pit basis under the existing colliery review procedure; and the principle of 'no compulsory redundancy' was to be maintained by improved redundancy terms, particularly for men under the age of 50, agreed with the government in January 1984, with £1000 per year of service plus other payments. This meant that a man aged 49, who had spent his working life in the industry, could receive a lump sum redundancy payment in excess of £36,000. These proposals were enthusiastically supported by the Secretary of State, (Peter Walker) and were scheduled to become effective from 1 April 1984 (Smith p.15). On this basis, it was hoped to close twenty uneconomic pits and reduce annual capacity by 4m. tonnes in 1984/5 (Thatcher p.343).

However, a policy of 'closure by stealth' on the basis of local agreement was becoming increasingly difficult to sustain as the industrial relations situation in the industry rapidly deteriorated. The overtime ban was enforced at all collieries, and increasingly began to impact on operations. Although little coal was then being produced in overtime, and most mineworkers worked little overtime, the ban had a disruptive effect on repair and maintenance work, and led to a situation when essential work, usually carried out at weekends on an overtime basis, had to be completed during the normal working week, thereby interfering with coal production. In turn this led to lay-offs

and loss of earnings, and local disputes because of this. Yorkshire led
the slide into chaos. By the beginning of March 1984, six Yorkshire
collieries were on strike, and on 28 February, the NUM South
Yorkshire panel called a strike of all pits in the NCB South Yorkshire
area from 5 March. This decision coincided with the announcement
by the Board Area Director (George Hayes) on 1 March that Cortonwood colliery was to close shortly, subject to discussions under the
colliery review procedure. (Similar discussions were proceeding in other
NCB Areas to identify the best way to achieve the planned reduction
in capacity of 4m. tonnes in 1984/5). Under the circumstances, the
Cortonwood announcement seemed particularly provocative: indeed
the view gained ground that this was a deliberate act of provocation
by MacGregor in order to trigger the strike. There is no evidence for
this. Ned Smith, the NCB Industrial Relations Director, was categoric
that if Cortonwood was the trigger for the strike, this was inadvertent,
and was due mainly to a failure in internal NCB communications to
the Area Director (Smith pp. 24–25). As he has stated: 'The announcement of the proposed closure of Cortonwood was not a deliberate
provocative act by the NCB. It would not, in fact, have been presented
in the manner in which it was if the Area Director had been given the
advice it was intended he should get' (Smith p. 27). Moreover, even if
the Cortonwood closure announcement had not been made, it is highly
likely that some other 'casus belli' would have presented itself as the
chaos in South Yorkshire escalated.

Notwithstanding the Cortonwood announcement on 1 March, over
the weekend of 3/4 March, NUM branch meetings at a majority (8
out of 15) of South Yorkshire collieries voted not to join the area strike
which had been called for 5 March. But this became irrelevant. On 5
March, all but four pits in the South Yorkshire Area were prevented
from working by intimidatory pickets from other collieries in the Area.
On the same day, a Special Council Meeting of the Yorkshire NUM
(with representatives from all branches in Yorkshire) decided to call a
strike across the whole of Yorkshire from 9 March, on the authority of
the January 1981 ballot which had been 86 per cent in favour of strike
action 'in principle' over closures, and which had been mounted by
Scargill when he was still Yorkshire President over three years earlier.

On 6 March (by which time all the pits in South Yorkshire were
either on strike or had been closed by pickets), the NCB met the
unions at a pre-arranged meeting of the Coal Industry National
Consultative Council (CINCC). Ostensibly, the Board sought to outline
its plans for 1984/5, in order to bring supply and demand into balance
by reducing output by some 4m. tonnes – similar to the reduction in

1983/4. (The out-turn for 1983/4 involved the closure of fifteen collieries, with seven mergers, reducing the number of operating collieries by 21, and the labour force by 22,000). It had also been intended that the meeting would discuss an agenda for a possible tripartite meeting between government, unions and the NCB to secure an agreed policy for the industry. In that regard, the Board stated that it had no intention of running the industry down; rather it hoped to maintain deep-mine capacity of at least 100m. tonnes (*NCB Annual Report, 1983/4* pp.6–7). But Scargill was not interested in the supply/demand balance, still less in discussions with the government on general policy statements. What he wanted from the NCB was a specific admission on the scale of closures and manpower losses planned for 1984/5. MacGregor was in no mood to water down his plans as a conciliatory gesture to the unions, and Cowan (the NCB Deputy Chairman) admitted that the manpower reduction in 1984/5 would be similar to that in 1983/4. The meeting of CINCC broke up in total disagreement (Routledge pp.142–3).

The National Executive Committee (NEC) of the NUM met on 8 March. Under 'Rule 41' the Scottish NUM asked for national backing for a strike already under way over the closure of Polmaise colliery, and Yorkshire for the strike being mounted over the closure of Cortonwood. Roy Ottey, the 'moderate' leader of the NUM 'Power Group' later recalled: 'Straightaway I smelled a rat. I realised the Yorkshire and Scottish calls could represent the start of a national strike without a ballot. It was obvious that Britain's miners were in danger of being brought out through mass picketing and intimidation' (Roy Ottey: *The Strike*, pp 60–61 quoted Routledge p.143). The NEC meeting declared the Yorkshire and Scottish strikes to be official, and that any similar action elsewhere would also be deemed to be official.

The small Kent coalfield immediately joined the strike. Although a majority of NUM branches in South Wales, and a significant number in the North East and Scotland decided against joining, by 15 March all collieries in these coalfields were either on strike or 'picketed out'. The traditionally 'moderate' areas held strike ballots which were heavily against strike action (see Table 2.13). North Derbyshire and Northumberland supported the strike with votes of 50 per cent and 52 per cent respectively.

From the evidence of the area ballots that were held, together with indications from the individual colliery branch votes in those coalfields where area ballots were not held, it is highly probable that Scargill and his colleagues would once again have been unable to secure the 55 per cent majority then required for a national strike if a national

Table 2.13: Strike Ballot Results in Traditionally 'Moderate' Areas. March 1984.
Per Cent

	For	Against
Leicestershire	11	89
Midlands	27	73
North West (Lancs)	41	59
North Wales	32	68
Nottinghamshire	27	73
South Derbyshire	16	84

Source: *NCB Annual Report 1983/4*

ballot had been called. For Scargill, therefore, the strategy of pro-ceeding by other means appeared to be justified, given that his objective was a national strike against the government. As Mick McGahey, the NUM Vice President had said: 'We will not be constitutionalised out of action' (Routledge p.144). From the beginning, the 'other means' meant primarily intimidatory mass picketing, often with 'flying pickets' (usually from Yorkshire) being sent to other areas to stop their collieries working. The level of picketing was such that many collieries joined the strike while still awaiting their ballot results. At pits where the great majority wished to work, large numbers of police were often required to overcome mass picketing.

By the end of March 1984, the battlelines had become clearer. Nottinghamshire, Leicestershire, South Derbyshire, Warwickshire and Cumbria continued to work, despite heavy picketing (and all opencast regions, which had minimal NUM membership, also continued to work). Lancashire, North Wales and Staffordshire were divided into striking and working pits. Scotland, Northumberland and Durham, Yorkshire, North Derbyshire, South Wales and Kent were fully on strike. The collieries which continued to work represented about a third of the NCB's deep-mined capacity, although the overtime ban continued to be observed at working pits, so that their output was less than normal. In spite of the absence of a national ballot, by invoking Rule 41 and, above all, by the use of large-scale (and illegal) mass picketing, Scargill had managed to launch a strike covering about two-thirds of the industry, but by means so divisive as to make a fully national strike very difficult, if not impossible to achieve.

Indeed, at the end of March, the 'moderates' on the NUM National Executive made a final attempt to get a national ballot called, believing even at that late stage that this would result in the strike (or, more accurately, strikes) being called off. Accordingly, at the meeting of the

National Executive on 12 April, they came with a motion for a national ballot of the membership. Ned Smith was of the view that this would have been carried by 14 votes to 10. But Scargill, as President, ruled the motion out of order, and it was never put to the vote. Instead, a national delegate conference (traditionally dominated by 'militants') was summoned, which summarily rejected the call for a ballot, and also changed the rules on any future ballots to provide for simple majority decisions for strike action (instead of the 55 per cent previously needed) (Smith pp. 73–74).

Picketing on a large scale continued. In some cases over 5000 'flying pickets' would arrive at selected Nottinghamshire collieries, which were kept open only by massive police presence. By early May, these tactics led to an increase in men on strike in Lancashire and Staffordshire, but this proved to be the last significant extension of the strike. Thereafter, the position stabilised until the late autumn, both in terms of the numbers of NUM members at work, and the numbers of pits producing coal.

The NUM's Road to Defeat

From May, there had been an increased emphasis in the mass picketing towards the users of coal, particularly BSC works at Orgreave in Yorkshire, Ravenscraig in Scotland, and Llanwern in South Wales. Attempts by thousands of miners to stop lorries taking coke from Orgreave coke ovens to Scunthorpe steelworks resulted in pitched battles with police ('The Battle of Orgreave'), accompanied by much violence and injury on both sides. But those tactics brought little benefit to the strikers, particularly as other groups of workers (other than some railway workers) showed little inclination to assist the NUM, in large measure because the strike had not been preceded by a national ballot.

During this time, there were negotiations between the NCB and the NUM, on the initiative of the NCB, to find a way of ending the strike. Discussions in May and June ended in deadlock, but were resumed in July when an attempt was made to draft a basis of a settlement on the issue of closures. Once again, the talks broke down when the NUM could not agree that pits could be closed on economic grounds. The use of the euphemism to describe uneconomic collieries as 'pits which had reserves which could not be beneficially developed' was unable to break the point of principle: Scargill would not be seen agreeing to 'economic' closures and MacGregor could not surrender the management prerogative to make such decisions. Further attempts were made in September and October (in the latter case under the auspices of ACAS) to find the basis for a settlement, in particular to find acceptable wording on the issue of economic closures, but without success. Ned Smith recalls how, on several occasions, the NCB and NUM appeared to be tantalisingly close to agreeing a form of words. But in any event, there must be serious doubts as to whether any 'settlement' could have endured after the strike, particularly if it had included any kind of undertaking by the Board to maintain deep-mined output at around 100m. tonnes a year. Such a target, to which the Board referred several times during the negotiations, would have been quite unrealistic without wide-ranging government guarantees of long-term market and financial support, which it is inconceivable a Thatcher government would have granted at that point.

In the meantime, the main threat to the NCB's position came from the National Association of Colliery Overmen, Deputies and Shotfirers (NACODS), the officials union, whose members played an essential part, not only in front-line supervision, but also, more crucially, in the regular safety checks required on a shift and daily basis by the Mines and Quarries Act. In April 1984, NACODS members had voted 56:44 in favour of strike action over closures, but this fell short of the two-thirds majority required by that union's constitution. However, under the special circumstances of the NUM strike, many NACODS members were placed in great difficulty. At the collieries which continued working, NACODS members crossed picket lines, together with the workers they supervised; but in the striking areas, pickets sought to prevent the officials from going to work. (Where they succeeded, safety checks at the non-working pits had to be carried out by management grades). The NCB sought to establish new guidelines whereby some NACODS members would be obliged to cross picket lines even at striking pits. This proposal gave rise to strong hostility, and in September a NACODS ballot brought an 82 per cent vote in favour of a strike over the new guidelines and the issue of closures. This was a major crisis, both for the NCB and the government, as it threatened the ability of the working pits in Nottinghamshire and elsewhere to continue to produce coal. An all-out strike by NACODS would have given Scargill the nationwide stoppage that he had been unable to secure on his own. For a time, the government was in a state of near panic. MacGregor was summoned to Downing Street. Mrs Thatcher is reputed to have told him 'You have to realise that the fate of this government is in your hands, Mr MacGregor. You have got to solve this problem' (quoted in Routledge p.177). The NCB withdrew their controversial guidelines on crossing picket lines, and, in order to address the closure question, proposed a revision of the industry's Colliery Review Procedure (CRP), which had been the vehicle for closures under the policy of 'stealth'. NACODS had raised the need to introduce binding arbitration in the last stages of this procedure, to avoid a situation where the Board was seen to be 'judge and jury' in the final closure decision. While the NCB, on legal advice, was unable to accept binding arbitration, it proposed a new final stage in the CRP, which would allow an independent tribunal to express an opinion on any matter referred to them after a national appeal. After talks at ACAS, this proposal proved acceptable to NACODS, and the strike threat was called off on 24 October 1984.

The resolution of the NACODS problem was the beginning of the end for the NUM strike. There was increasing evidence from November onwards that the government was giving signals (often through unofficial channels) that it was no longer interested in any settlement which did not include explicit NUM agreement to the closure of uneconomic pits (Smith pp.149–57). This was tantamount to a demand for unconditional surrender. The drift back to work increased: over 10,000 men returned to work in November, the number of pits without any men regularly at work halved from 84 to 41, and coal production was able to begin at more pits, although this process was accompanied by much violence on the picket lines. As the drift back to work began to accelerate in February, an attempt at a settlement, brokered by the TUC after discussions with government, and embodying the revised CRP agreed with NACODS, was rejected by the NUM Executive. By 27 February 1985 the number of NUM members not on strike passed 50 per cent. An NUM Delegate Conference on 3 March by a narrow majority (98 to 91) voted for a general return to work on 5 March, but without a settlement. By the first week of April 1985, the overtime ban was called off, and the NUM had accepted the wage offer outstanding from 1983 and agreed to consider a similar offer for 1984. The Great Strike was over.

Why the Great Strike was Defeated

Edward Heath, who as Prime Minister had been willing to see the miners as a 'special case', and who had particular reasons to feel bitterness towards the NUM, neverthless regarded the Great Strike and its outcome as tragic 'for those that believe in One Nation', and observed that 'There was an unpleasant atmosphere of triumphalism on the Conservative benches at this time' (Heath p. 586).

However, the Thatcher government saw no reason not to regard the defeat of the strike as a great triumph. Later Nigel Lawson wrote that 'Just as the victory in the Falklands War exorcised the humiliation of Suez, so the eventual defeat of the NUM etched in the public mind the end of militant trade unionism which had wrecked the economy and twice played a major part in driving elected governments from office' (Lawson p.161). Margaret Thatcher was even more forthright: 'What the strike's defeat established was that Britain could not be made ungovernable by the Fascist Left. Marxists wanted to defy the law of the land in order to defy the laws of economics. They failed, and in doing so demonstrated just how mutually dependent the free economy and a free society really are. It is a lesson no one should forget' (Thatcher p. 378).

The clarity with which the Thatcher government saw that the defeat of the Great Strike opened up the possibility of irreversibly breaking the power of the NUM (the cornerstone of the government's 'political agenda' for the coal industry), has lead many to believe that the government itself by careful planning deliberately engineered the outbreak of the strike in early 1984, at a time of year unfavourable to the union, when the winter was nearly over, thereby ensuring the NUM's ultimate defeat. This view is unconvincing for two reasons in particular.

In the first place, there is a more credible alternative explanation. As we have seen, it was very clear that Scargill and his close associates (particularly in his power base in Yorkshire) were determined to mount a national strike against the government, although in the event the timing was inadvertent and opportunistic. Secondly, the government could not have foreseen the way in which the chaos in early March 1984 would result in the strike beginning when it did, and with a third of the industry continuing to work.

On the other hand, the Government had come to believe that a challenge by Scargill was almost inevitable at some point. Moreover, the NUM leadership was making 'impossibilist' demands that the coal industry must be greatly expanded, that coal must be produced whatever the cost, and that the NUM should have power of veto over NCB decisions. This position was not remotely negotiable. In these circumstances, the Government sought to take such precautionary measures as they could in the event of an NUM strike occurring. Nigel Lawson, who was Energy Secretary from 1981 until June 1983, claims some credit for the opposition of 'moderate' Midlands miners to Scargill, in that he persuaded Mrs Thatcher and Michael Heseltine to agree that one of the NCB's proposed new mines in the Vale of

Belvoir (Asfordby) should proceed, notwithstanding environmental objections (Lawson pp.145–6). But this was a subsidiary point. As Lawson put it 'The overriding need in preparing for a coal strike was to increase substantially power station endurance – that is, the length of time the power stations could continue to meet the nation's needs in the event of a complete cessation of coal production in the UK.' Accordingly in September 1981 he asked the CEGB to produce 'a detailed and costed plan for increasing power station coal stocks, laying in extra supplies of the ancilliary chemicals required in electricity generation, transporting coal by road rather than rail, and for conserving coal by increasing the amount of electricity generated by oil-fired power stations' (Lawson p. 149). The plan was approved by Mrs Thatcher in February 1982 (Lawson p. 150). This policy was certainly facilitated by the fact that, with the NCB continuing to over-produce, colliery stockyards in many cases were full (stocks having reached 25m. tonnes by March 1982), so that it was difficult to find anywhere else for the surplus output to go. Nevertheless, the CEGB initially resisted the build-up of coal stocks at power stations, and it required the personal intervention of Mrs Thatcher to ensure that the CEGB followed what was a clearly stated, if secret, government policy (Personal Communication: M.J. Edwards). In the event, the power station coal stocks increased by some 15m. tonnes between March 1982 and December 1983.

These higher levels of power station coal stocks gave the government sufficient assurance to stand firm in March 1984, and to avoid making damaging concessions to try and stop the strike spreading (as they had done in February 1981). Nevertheless, it is unconvincing to argue that the increase in power station stocks over the two years before the Great Strike of itself proves that the government had a settled plan to 'take on the miners' in 1984, given the great risks and uncertainties attendant on such an enterprise. The higher stocks established by early 1984 did not of themselves guarantee government victory over the miners. There was no means of knowing before the event that Nottinghamshire and the other 'moderate' coalfields would continue to work throughout the strike, that any difficulties with NACODS would be so readily solved, or that the support from other trades unions would be so limited. Indeed, Mrs Thatcher was uneasy when in April 1984 the NUM changed the rule on the majority required for a strike in a national ballot from 55 to 50 per cent, since this suggested that Scargill might at some stage wish to invoke such a rule. As she said in her memoirs: 'Would a ballot held during the strike, with emotions raised, produce a majority for or against Mr Scargill? I was not entirely sure' (Thatcher p.349). In the light of these risks, to have deliberately engineered the outbreak of the strike would have been manifest folly, given the disastrous consequences to the government if such a strike had proved to be successful. Indeed, in the six months or so before the strike, the government appeared to be working primarily on the basis that 'defeating Scargill' meant once again denying him a strike by ensuring that there would not be a successful national ballot for industrial action, thereby enabling the NCB policy of decentralised one-by-one closures to continue. It was for this reason that the enhanced redundancy terms were agreed in January 1984, and, according to Ned Smith (the NCB Industrial Relations Director), Peter Walker (The Energy Secretary) told MacGregor at a meeting in February 1984 that it was the government's wish that no precipitate action which might give rise to a strike was to be taken, either at national or Area level (Smith p.24).

With hindsight, Lawson exaggerates the extent to which his policy of building up power station stocks was responsible for providing the necessary 'endurance' to out-stay the twelve month strike. When he says that power station stocks rose to 58m. tonnes in December 1983 (Lawson p. 150) this should refer to *total* stocks, over 40 per cent of which were still at collieries and opencast sites. And in terms of statistical analysis, it can be shown that the withdrawal of power station coal stocks during the strike played a relatively subsidiary role in ensuring that electricity supplies were maintained (although the high level of stocks played a crucial role in maintaining the public's belief in the ability of the government to defeat the strike).

The contribution of the various factors which contributed to meeting UK energy needs during the period of the Great Strike can be analysed conveniently in terms of the outcome for the financial year 1984/5 (which, within two or three weeks, coincided with the strike) as compared with 1983/4. (See Table 2.14)

In spite of the massive disruption to coal supply arising from the strike, coal's share of the primary energy market, excluding power stations, showed little change. But in the power station market, there was a huge swing of nearly 40m. tonnes from coal to fuel oil, undertaken without regard to the enormous cost, which accounted for the greater part of the reduction in total coal use between the two years.

But between 1983/4 and 1984/5, there were also significant changes in the way in which the UK demand for coal was met; with the huge reduction of 60m. tonnes in deep-mined output being only partially offset by more imports and stock lift. (See Table 2.15)

Table 2.14: UK Primary Energy Consumption 1983/4 and 1984/5. Million Tonnes Coal Equivalent

	Power Stations 1983/4	1984/5		Other Markets 1983/4	1984/5		All Markets 1983/4	1984/5	
Coal	82	43	-39	30	24	-6	112	67	-45
Oil	9	47	+38	98	99	+1	107	146	+39
Nuclear	16	19	+3	2	2	-	18	21	+3
Gas	-	1	+1	77	78	+1	77	79	+2
Other	2	1	-1	-	-	-	2	1	-1
	109	111	+2	207	203	-4	316	314	-2

Source: *NCB Annual Report 1984/5*

Table 2.15: UK Coal Supply and Demand 1983/4 and 1984/5. Million Tonnes

	1983/4 *	1984/5
UK coal consumption		
Power Stations	82	43
Other Markets	30	24
Total	112	67
Demand met by		
NCB Deep mined	88	28
NCB Opencast	14	14
Other UK production	5	2
	107	44
Exports	- 7	- 1
Imports	5	11
Stock lift		
Distributed	4	11
Undistributed	3	2
Total	112	67

*   53 weeks adjusted to 52 weeks.

Source: Derived from *NCB Annual Report 1984/5*

If we assume that, in the absence of the strike, UK coal consumption in 1984/5 would have been 112m. tonnes (i.e. the same as in 1983/4), then the relative contribution of the various factors to meeting that demand can be broadly assessed as follows. (Table 2.16)

Table 2.16: How Underlying Coal Demand was Met During the Great Strike.
Million Tonnes

Continued working of the 'moderate' coalfields, and of opencast sites	42
Switch from coal to fuel oil in power stations	38
Net stock lift	13
Net imports	10
All other factors	9
Total	112

The ability of the country to survive the strike owed much to the endurance of those miners who continued to work, the failure of the NUM to target and prevent opencast coal production and the considerable feats of improvisation by the management of the NCB and the CEGB. But the support of the government was crucial. Although the government did not engineer the outbreak of the strike, once it had started with widespread and illegal violence, the government had little option but to take measures to ensure that the strike did not succeed.

Throughout the strike, the NCB was in virtually daily contact with the Secretary of State (Peter Walker) and he in turn was in constant touch with the Prime Minister on the progress and handling of the strike. Mrs Thatcher sought to determine the appropriate stance for the government to take: 'I repeatedly made it clear that prime responsibility for dealing with the strike lay with the managements of the NCB and those other nationalised industries involved'; but 'so much was at stake that no responsible government could take a "hands-off" attitude: the dispute threatened the country's economic survival. Consequently, I tried to combine respect for their freedom of action with clear signals as to what would or would not be financially and politically acceptable.' (Thatcher p.347) So far as the NCB was concerned these 'signals' certainly included the need to avoid a 'capitulation' settlement, and an assurance that the government would, at the end of the day, underwrite the huge financial losses involved in ensuring that the strike did not prevail. Above all, the CEGB were encouraged to disregard the huge additional cost of using fuel oil at their under-utilised oil stations, and at their coal stations, to replace about half of their coal consumption, or the cost of road transport to take coal from working pits to power stations. As Nigel Lawson said later: 'It was essential that the government spent whatever was necessary to defeat Arthur Scargill' (Lawson pp.160–61).

The government also had to give very careful consideration to the issues of law and order, and to its relationship with the police and the courts. The Attorney-General made a written statement to the House of Commons, setting out the legal powers of the police to deal with mass picketing, including the power to turn back pickets on their way to the picket line when it appeared that a breach of the peace was threatened. The government also made it clear that it would support the police in the exercise of these powers, which were common law powers under the criminal law, and predated the government trade union legislation (Thatcher p.348). However, the government also had much internal debate on whether the nationalised industries involved in the strike should seek civil remedies against the NUM which, if successful, would have limited the NUM's ability to finance mass pickets; but finally decided that this would be counter-productive (Thatcher pp. 353–4).

Yet even with this well-considered blend of prudence and resolution on the part of the government, it is difficult to see that the strike could have been endured if there had been widespread outside support for the strike. On 28 March 1984, Scargill wrote in the *Morning Star* that 'the NUM is engaged in a social and industrial Battle of Britain ... what is urgently needed is the rapid and total mobilization of the Trade Union and Labour movement'. But this did not occur. As the NCB reported (*1984/5 Annual Report*, p.4), two national dock strikes called in July and August were short-lived, and 'by the time of the Trades Union Congress (TUC) conference in September it was clear the NUM would not receive any significant outside support from workers other than rail', largely because of the absence of a pre-strike ballot of the whole mining workforce. What was true of the TUC applied with even greater force to public opinion, due in large measure to the actions and image of Arthur Scargill. Roy Hattersley, who has written from a left-of-centre perspective, and who has rightly drawn attention to the great suffering of the striking miners and their families over the year of the strike, observes that Scargill's 'unpopularity among the general public was so great that it more than offset the nation's traditional sympathy for the miners' (Hattersley p.295).

It is difficult to resist the analogy of war. Once the Great Strike was well underway, economic cost/benefit calculations became irrelevant. (It is estimated that the strike added £2.7 billion to the Public Sector Borrowing Requirement in 1984/5). For each side, as the months went by, increasingly the only language was of victory or defeat. The victors, as Mrs Thatcher proclaimed, were the government. The defeated were Arthur Scargill and the NUM. The victims were the striking miners and their families. For the industry itself, the results were more complex.

CHAPTER 3
'TURNING BRITISH COAL INTO A BUSINESS'?

A New Start for the Coal Industry?

In the months following the end of the Great Strike, the emphasis of both the NCB and the government was on a new start for the industry. Ian MacGregor wrote in his Chairman's statement for the Board's 1985/6 Annual Report of 'establishing a high-volume, fully viable industry able to meet the competition and make a positive contribution to the national economy'; and, symbolic of the new opportunities 'in which we can put our problems of the past behind us, the Board have decided to trade under the name of British Coal' (Note: the legal entity remained the National Coal Board, until renamed the British Coal Corporation in the Coal Industry Act 1987. For ease of reference, we refer from now on to British Coal or 'BC').

The industry showed great resilience, with a rapid recovery of deep-mined output after the strike, in spite of the damage to underground roadways and coal face equipment caused by long periods of inactivity during the strike at the majority of pits. At the same time, 1985/6 saw the beginnings of the programme of radical restructuring, which resulted in a deep-mined output some 14m. tonnes below the level before the strike, and the attainment of record levels of productivity, with the closure of 36 high-cost collieries and a reduction of 33,000 in the colliery labour force. (See Table 3.1)

Table 3.1: Trends in Deep-mined Output and Productivity 1983/4 to 1985/6

		1983/4	1984/5	1985/6
	Deep-mined output (m.t.)	90.1	27.6	88.4
Plus	Tonnage lost through disputes (m.t.)	13.1	67.2	1.0
	Underlying trend (m.t.)	103.2	94.8	89.4
	No. collieries at year end	170	169	133
	Average manpower ('000)	191.5	175.4	154.6
	Net manpower change ('000)	-21.3	-9.7	-32.9
	Output per man year	470	157	571

Source: *NCB Annual Reports*

Coal demand also recovered rapidly. UK coal consumption, both at power stations and overall, was higher in 1985/6 than at any time since 1980/1 (much influenced by a sharp increase in electricity demand). Oil prices were still high, and consumers, particuarly power stations, switched back to coal as soon as possible. BC sales also benefited from the measures to rebuild power station coal stocks after the strike. During 1985/6, BC lifted 11m. tonnes of stock, most of which was for this purpose.

By March 1986, BC's stocks stood at only 8.3m. tonnes, the lowest year-end level since March 1974. Even if the rebuild of power station stocks in 1985/6 is disregarded, the supply and demand for BC output were in balance. The endemic over-production of the years leading up to the Great Strike had been resolved by a combination of demand recovery and reductions in deep-mined output through the closure of high-cost pits. (See Table 3.2)

Table 3.2: Coal Supply and Demand 1985/6. Million Tonnes

	UK coal consumption	
	Power stations	86
	Other markets	32
	Total	118
Plus	rebuild of consumers' stocks (mainly power stations)	9
	Total UK coal demand	127
Met by	BC production (Deep mines, Opencast etc)	104
	BC stock lift	11
	Total BC sales	115
	Other UK production	3
	Less BC exports	(3)
	Imports	12
		127

Source: *NCB Annual Report 1985/6*

The financial results for 1984/5 were wholly distorted by the effects of the Great Strike. Deep mines had a collective operating loss of £1.673m. (including a provision of £340m. for the likely costs of recovery at collieries which had been on strike). The NCB's overall loss of £2,225m. was covered by a government deficit grant payable partly under the Coal Industry Act 1983 and partly under a new Coal Industry Act 1985. The External Financing Requirement of £1,720m. was met by temporary borrowing from government. These financial arrangements enabled a new start to be made in 1985/6, although the large debt burden still remained.

The Board's financial position in 1985/6 showed considerable improvement as compared with 1982/3 (the last financial year unaffected by the Great Strike). There was an operating profit of £601m., including £232m. at deep mines; and the small net deficit of £50m. was more than accounted for by terminal depreciation and additional social costs arising from closures. (See Table 3.3)

Table 3.3:   NCB Financial Results 1982/3 and 1985/6. £Million

	1982/3	*1985/6*
Operating Profit/(Loss)		
Deep mines	(317)	232
Opencast	192	343
Other activities	26	26
Total	(99)	601
Sales of fixed assets	20	24
Interest	(364)	(437)
Social costs less grants	(49)	(170)
Terminal depreciation	-	(66)
Miscellaneous items	7	(2)
Deficit before deficit grant	(485)	(50)

Source: *BCC Annual Report 1986/7:* including changes in accounting treatment of earlier years

To a considerable extent, the operating profit of deep mines in 1985/6 reflected the release of the £340m. provision made in 1984/5 for strike recovery costs. Although precise measurement of the actual level of these costs is not possible, it would appear that the provision was over-generous, so that the deep-mined operating profit in 1985/6 was inflated. Nevertheless, average operating costs (before deducting the provision) were some 5 per cent below the 1982/3 level in 'real' terms. The prospects for further improvements were enhanced by the effects of the defeat of the Great Strike, which not only eroded the ability of the NUM to impede the closure of uneconomic collieries, but also gave management at local level more freedom of action to exploit the potential productivity improvements becoming available from improved mining technology, particularly at the coal face and in roadway drivage.

Against this background of the elimination of over-production, and rapidly improving deep-mined performance, it was not unreasonable to assume that a fundamental recovery in the industry's fortunes was taking place, and that it was realistic, to use the phrase of Ken Moses (then the BC Technical Director), seriously to begin the process of 'turning an institution into a business'.

But what did this entail? The objectives agreed between the government and Ian MacGregor, shortly after he became Chairman in September 1983, were similar to those agreed with the previous Chairman, Norman Siddall; but there was for the first time a reference to profit maximisation: British Coal should (inter alia) seek to 'maximise its long-term profitability by securing those sales which are profitable on a continuing basis'. Such a formulation was somewhat obscure: what were 'profitable sales'? What was the level of aggregation, and to what extent were *average* revenues and costs to be used? These objectives did not make explicit whether a frontal assault was to be made on the endemic cross-subsidies within the industry; but if BC was to operate as a fully commercial business, this issue would ultimately have to be faced. 'Profit maximisation' would mean that BC ought not to plan for continued operations at any individual colliery which had no real prospect of profitable operation under competitive conditions. This would have been a radical departure from traditional coal industry thinking. The view was widely held, even in the highest levels of BC management, that the Coal Industry Nationalisation Act (1946) required the industry 'only' to break-even (whereas this was in fact a minimum requirement taking good years with bad); so that the industry was in some sense 'allowed' as much loss-making output as profitable output. A scarcely more sophisticated variant of this view was that deep mines needed to make an aggregate operating profit only to the extent to which open cast and other activities were unable to cover BC's interest charges. Such views were very much at variance with making BC a fully commercial business, and were closely linked to the assumption that BC's real objective (whatever the government might say) was to maintain the size of the industry at the highest possible level, subject to certain overall financial constraints.

It was to try to establish a radical break from this thinking that Ken Moses and the author sought to devise new economic criteria to be applied to individual collieries. This would also break the link with output targets, which the industry's management was reluctant to abandon. The result was a document called 'A New Strategy for Coal', which was endorsed by the Board in late summer 1985. This relinquished specific targets for sales and production, and instead introduced 'competitive cost' criteria to judge whether collieries should be kept in production, and to justify further investment. These cost criteria were also designed to give 'minimum regret' decisions on closures and investment. The three critical cost levels were:-

- £1.65/GJ would be the upper limit of acceptable cost (except for the small number of collieries producing anthracite or other specialist coals, mainly in South Wales). This meant that collieries with no prospect of producing at less than this level of cost would be early candidates for closure.
- £1.50/GJ would be the maximum level of operating cost for a colliery to have an assured long-life future, and so be worthy of major capital investment.
- £1.00/GJ should be the cost of 'incremental' output, if investment were to be made at a colliery in order to increase output.

These levels were related to the price at Rotterdam of internationally traded steam coal in 1984 of about £1.40/GJ and after allowing for transport costs. 'Incremental' production was valued at 'net-back' realisations in the export market, which was BC's marginal market.

The Oil Price Collapse: The World Changes

Yet this neat logic, and the feeling of a new beginning which was starting to pervade the upper echelons of BC were overtaken by a fundamental change in world energy markets: the collapse of crude oil prices. Since the second oil 'shock' of 1979/80, new market forces had emerged which in due course were to erode the power of OPEC to sustain very high oil prices. The events of 1979/80 depressed the demand for crude oil, both through adverse macroeconomic effects on OECD countries, and by encouraging additional substitution of oil (particularly heavy fuel oil) by coal, nuclear power and natural gas; and at the same time the incentives to develop non-OPEC crude oil sources were increased. Strains began to show within the ranks of OPEC. By the end of 1982 only Saudi Arabia was selling all its oil at the 'official' OPEC price of $34 per barrel; other members were involved in discounting. More than half of the internationally traded crude oil was either on the spot market or sold at prices linked to the spot market (Daniel Yergin: *The Prize*, p.722). In 1983, OPEC reduced its 'official' price to $29 per barrel with new production quotas for individual members designed to balance supply and demand for OPEC oil. But also in that year, the New York Mercantile Exchange (NYMEX) introduced a futures contract in crude oil: an act of great significance, and a portent of the radical changes taking place in the world oil market. In 1984, there was increasing evidence that OPEC's new quota system was not working, with systematic cheating by countries wishing to defend their market share. It was no longer possible for OPEC to defend both price and volume, and in 1985 the Saudis moved from a defence of price to a defence of volume. The mechanism for this change of policy was 'net-back pricing' contracts, whereby Saudi Arabia would be paid on the basis of what refined oil products actually earned in the market place, less an agreed margin for refiners. This arrangement triggered other 'net-back' deals. Soon there was no official Saudi price, and could be no official OPEC price. In November 1985, the communiqué of the OPEC meeting made it clear that the objective was no longer the defence of a particular level of crude oil prices but to 'secure and defend for OPEC a fair share in the world oil market …' (Yergin pp.748–50). This was a perfect recipe for a price collapse, which duly occurred in early 1986, when prices more than halved.

Nor was this a temporary phenomenon. In the three years following the price collapse, the price of crude oil and heavy fuel oil (the oil product of greatest relevance to the coal industry) was less than half the level which obtained in the three years prior to the Great Strike, after adjusting for inflation. Of greater direct significance to the UK coal industry, the trend in internationally traded steam coal prices followed a similar path. Although the development of this trade had been characterised by coal-on-coal competition, which had been the main determinant of prices, the price of fuel oil had always been seen as operating as a ceiling for steam coal prices, in view of the inter-changeability of coal and fuel oil in much of the West European power station market. The very large fall in oil prices thus put further downward pressure on international coal prices. By 1987, the c.i.f. contract price of steam coal imported into the EU for power stations was, in 'real' sterling terms, only half the price level preceding the Great Strike.

The oil price collapse and the rapid fall in international coal prices were of great significance in the conduct of energy policy so far as it concerned the UK coal industry. The rapid erosion of the power of OPEC and of the NUM within a short period of time encouraged the view that British Coal was to be regarded as a business rather than a strategically essential industry.

In March 1987, the Secretary of State for Energy settled objectives for the industry with the new British Coal Chairman, Sir Robert Haslam (who had succeeded Ian MacGregor in September 1986). The main objectives were as follows:

> Coal production, like any other business, must earn a satisfactory return on capital while competing in the market place. The basic objective of British Coal must be to earn a satisfactory rate of return on its net assets and achieve full viability without government support. British Coal should accordingly aim to improve its profitability so as to achieve, after payment of interest and accrual of social grants, breakeven for the year 1988/9 as a whole ... and thereafter generate an increasing surplus on revenue account and to contribute increasingly to self-financing.

> British Coal should aim to maximise its long-term profitability by concentrating on low-cost production, and on those sales which maximise profit on a continuing basis in competition with other fuels. It should plan its marketing, production, capital investment and research and development accordingly, bring productive capacity into line with its continuing share of the market, and ensure an adequate return on new capital investment in accordance with the principles set out in the White Paper 'The Nationalised Industries' (Cmnd 7131).

> British Coal's objective should be to ensure that over the period to 1989/90 the operating costs of its mining activities per gigajoule of coal produced are reduced in real terms by at least 20 per cent, compared with the level recorded in 1985/6.

Although these objectives were not dissimilar to those agreed with the two previous Chairmen (Norman Siddall and Ian MacGregor), they were specific in terms of cost reductions to be sought, and the emphasis on maximising long-term profitability was greater.

Sir Robert Haslam saw this as a time of opportunity. In his 1986 Christmas message to staff, he spoke of the 'remarkable recovery since the strike. Productivity is the highest in the industry's history; collieries are regularly breaking records and costs are being reduced. On the other hand, we have been faced with the most formidable of all challenges – the collapse of international oil prices, the consequent pressures on our own prices, and the ever present threat of cheap imported coal.' Nevertheless, he felt able to say that 'we are about to enter a New Year (1987) in which I hope we sail into calmer waters. The industry's major dramatic restructuring is nearing completion, but of course, healthy restructuring will always be with us. We are sustaining investment at some £650 million a year. I hope we are set for a period of stability in which we can all get on with the basic task of creating an efficient energy industry with a bright future for coming generations.'

In fact, these sentiments proved to be much too sanguine. The combination of a challenging financial framework and a deteriorating external business environment led to

- government-supported management of the market for UK coal, both in terms of volume and price, primarily through the relationship with the electricity supply industry.
- continuing radical restructuring in terms of colliery closures and manpower reductions.
- a downturn in capital investment, particularly for major projects.
- large-scale government support to facilitate the restructuring process.

We examine each of these in turn.

Management of the Market for UK Coal

In this period following the Great Strike, sales to power stations continued to represent the major part of British Coal sales, rising to 80 per cent by 1988/9. (See Table 3.4)

Table 3.4: BC Sales 1986/7–1989/90. Million Tonnes

	1986/7	*1987/8*	*1988/9*	*1989/90**
Power stations	79.0	80.0	78.8	75.8
Other markets	23.1	22.1	19.7	18.5
Total	102.1	102.1	98.5	94.3
Power stations as % of total sales	77%	78%	80%	80%

*   53 weeks

Source: *BC Annual Report 1994/5*

Moreover, coal and UK coal in particular, continued to dominate the total fuel supplies to UK power stations: over the four years 1986/7 to 1989/90 BC provided two-thirds of the fuel for UK power stations. (See Table 3.5)

Thus, although over the four-year period there was a slight fall in the dependence of the electricity industry on UK coal, and a small rise in the dependence of BC on this market, the position was basically

Table 3.5: UK Power Station Fuel Consumption 1986/7–1989/90. Million
Tonnes Coal Equivalent

	1986/7	1987/8	1988/9	1989/90
Coal	81.9	85.9	80.7	82.1
Nuclear	19.5	17.2	23.0	23.5
Oil	10.2	8.1	8.5	11.2
Natural gas	-	-	-	-
Hydro	2.0	1.8	2.3	2.0
Electricity imports	2.7	5.1	5.3	3.9
Total	116.3	118.1	119.8	122.7
Coal as % of total	70%	73%	67%	67%
UK coal as % of total	68%	68%	66%	62%

Source: *BC Annual Reports*

stable, with a high degree of interdependence, as had been the case in
the period before the Great Strike.

This broadly stable position was not the result of a normal
commercial relationship. Both the major generators (CEGB and SSEB)
were nationalised monopolies, and BC was a nationalised industry
operating in a highly politically-sensitive market. Nevertheless, both
the CEGB and BC had seen the benefits of a formal long-term
agreement governing volumes and price of coal supplier to power
stations. We have already noted that in 1979, the NCB (BC) and the
CEGB entered into a broad Understanding on coal supplies and prices
covering a five-year period. This provided that the CEGB would use
their best endeavours to take 75m. tonnes a year in return for the
NCB keeping their coal prices within the rate of UK inflation. In this
way, the CEGB gained assurance that the price of their largest source
of fuel would not increase in 'real' terms at a time when much opinion
was expecting further rises; and BC gained a guarantee on sales
volumes given the potential downsides on demand and possible
competition from other fuels, including international coal. The Understanding was not an enforceable contract, but the arrangement was
adhered to through the 1980s, with periodic modifications to reflect
market conditions.

The Understanding had been re-negotiated in 1983, incorporating
the concept of two-tier pricing. This adopted the broad principle that
in return for BC's willingness to keep its prices *below* UK inflation, and
to align its prices against internationally traded steam coal on that
tonnage of BC coal which it was practicable for CEGB to replace with

imports in the short term, it would be reasonable for BC to supply about 95 per cent of CEGB's coal requirements. This would allow a reasonable market share to small private UK mines, but would restrict imports to very low levels. The 'base tonnage' (typically 65m. tonnes a year) would be sold at a price which would be increased annually by only 85–95 per cent of the change in RPI. The remaining tonnage (typically 8m. tonnes a year) would be aligned to the delivered price of imported coal at South East power stations.

It was this Revised Understanding which came under pressure when oil prices collapsed in early 1986, and imported coal prices looked to follow suit. The CEGB raised with BC and the government the economic case for reverting to a high oil burn (MMC Report on the investment programme of the British Coal Corporation Cm.550 Appendix 2.2 para 9); and in evidence to the Energy Committee, the CEGB stated that a new arrangement was required since the object of the Understanding had been 'to expose the NCB to international competition at the margin'. They also concluded that their coal imports could be built up to a level of 30m. tonnes per annum over a period of 3–5 years, and that such a level of imports would save them £550 million per annum. Whatever the validity of that figure, clearly a new arrangement was required (Energy Committee: The Coal Industry (HC165) para. 27). The government was involved. In its response to the Energy Committee's report, the government stated that

> whilst it is important that the industries should be allowed to run their own affairs, they keep the government informed about their commercial negotiations, which obviously have major implications for public expenditure. ... The government does not believe it would be right to use the (generating) Boards' public sector status to protect British Coal from market forces ... The government encouraged British Coal and the CEGB to reach a commercial agreement covering the supply of coal which took due account of the fall in oil prices last year (Energy Committee HC387 paras 6.1, 6.2).

Nevertheless, the new arrangement which emerged was far from embodying the full exposure of BC to 'market forces'. Considerable weight was given to BC's concern that their programme of orderly restructuring should not be jeopardised, and the CEGB's interest in a viable coal industry and predictable prices (MMC Cm. 550 Appendix 2.2 paras 12 and 13). Indeed, the new Understanding was the result of the establishment of a common interest between the CEGB and BC, rather than of government influence. The government's initial recommendation at the time was that BC should make a short-term arrangement for the rest of 1986, rather than for five years. But Ian MacGregor decided that a five-year deal should be entered into, and the government acquiesced. (Private Communication: M. J. Edwards)

The new Understanding was agreed in June 1986, covering the five years from April 1986, and provided BC with the opportunity to sell at least 70m. tonnes a year to the CEGB. However, the tonnage to be supplied by BC to the CEGB was to be divided for pricing purposes into three tranches. The first (highest price) tranche was initially set at 50m. tonnes per annum, but was to be reduced progressively over the five years. The second, intermediate tranche would have a price reflecting prices of oil and imported coal, and a volume rising as the tonnage in the first tranche declined. The third (lowest price) tranche would align to the price of imports at coastal stations, and would cover 12m. tonnes, with 4m. tonnes from the third tranche earmarked for the 'Qualifying Industrial Customer Scheme' (QICS), designed to give lower priced electricity to large industrial consumers. Both the first and second tranche prices would be subject to escalation of 85–95 per cent of the annual change in RPI. The overall impact of the new terms of the Understanding over the first two years were estimated to represent a reduction of some £300m. per annum in the CEGB's coal costs, and hence in BC revenues (HC 165 para 28).

There were further modifications to the Understanding in 1988, when, to reflect market conditions, further 'real' price reductions were made across three-quarters of the CEGB's requirement (*BC Annual Report 1988/9* p.12), and these arrangements continued until the privatisation and restructuring of the CEGB in 1989/90.

Arrangements with the SSEB reflected local circumstances, but the concepts and principles were similar to those applied to the CEGB Understandings. British Coal's view of the importance of these Understandings was expressed in their note to the MMC, published in January 1989 (MMC Cm.550 Appendix 2.2 para 16). 'The Understandings have stood the practical test of major economic and political changes. They have provided both the Electricity Supply Industry and the British coal industry with a degree of stability and continuity on which to base their planning and operations whilst paying proper regard to changes in the energy market, sustainable in the medium term.' Indeed, it is difficult to exaggerate the importance of the Understandings with the CEGB (and the similar smaller arrangements in Scotland) in the overall management of the industry. Effectively, they provided a guaranteed outlet for some 80 per cent of total sales, and at predetermined prices, which were on average well above international market levels. On the other hand, the provision in the Understandings for falling prices in 'real' terms represented a progressive pressure to increase efficiency and reduce costs.

Non-power station markets were far less significant in determining the industry's total sales volume, but were on a declining trend. (See Table 3.6). BC sales to coke ovens continued to decline as a result of the closure of coke ovens and reducing availability of UK coking coal capable of meeting the increasingly stringent quality requirements of the steel industry. The domestic coal market was likewise in long-term decline, reflecting the increasing dominance of gas. In the industrial market, strong marketing efforts were made to at least retain sales, although this task was made more difficult by strong competition from oil, gas and imported coal, and by the government's decision to end from June 1987 the Coal Firing Grant Scheme (which had been introduced in 1981 to encourage the conversion of industrial boilers to coal). BC's response was to follow a 'fully commercial pricing policy' by aligning to the competition. While this enabled the volume of industrial business broadly to be stabilised, it was at considerable cost in terms of revenue. Exports were generally uneconomic, although a small presence was maintained (so that this market could be used more readily as a regulator for surplus production if required).

Table 3.6: BC Sales to Non-power Station Markets 1985/6–1989/90. Million Tonnes

	1985/6	1986/7	1987/8	1988/9	1989/90*
Coke Ovens	5.5	4.5	4.2	3.7	2.4
Domestic**	6.6	6.0	5.2	4.2	4.0
Industry and Other UK	10.7	10.4	10.5	10.0	9.5
Exports	3.3	2.2	2.2	1.8	2.6
Total	26.1	23.1	22.1	19.7	18.5

*   53 weeks
** Includes Manufactured Fuel Plants

Source: *BC Annual Report 1994/5*

Thus, in the period between the end of the Great Strike in 1985 and electricity privatisation in 1990, the main emphasis of the management of the UK coal industry was to maintain the volume at the highest level possible, mainly through the arrangement with the electricity generators, but also by 'commercial pricing' in the industrial market. Notwithstanding the wording of the Objectives given by the government, there was no sense in which BC was following a profit-maximising policy. Financial targets were fundamentally still treated as constraints, rather than as prime objectives. In practical terms, therefore, British Coal's objective could be restated in the following terms: *to maintain the industry (particularly the deep-mined industry) at the highest practicable level consistent with a reasonable balance between supply and demand and at least an overall break-even after interest and 'social cost' grants.*

In overall volume terms, this policy worked. The overall reduction in output and sales by 1989/90 was relatively modest and controlled, and a broadly stable balance between supply and demand was achieved. (See Table 3.7)

Table 3.7:   BC Output and Stocks 1986/7 to 1989/90. Million Tonnes

	1986/7	*1987/8*	*1988/9*	*1989/90**
Output				
Deep mines	88.0	82.4	85.0	75.6
Opencast	13.3	15.1	16.8	17.5
Licensed	2.0	2.1	2.1	2.1
Total	103.3	99.6	103.9	95.2
Year end stocks	9.8	5.8	9.7	9.1

*   53 weeks

Source: *BC Annual Report 1994/5*

Two economic consequences followed from this approach. First, the savings accruing from the targeted 20 per cent 'real' reduction in average cost per tonne, between 1985/6 and 1989/90, provided for in the government's objectives published in March 1987, were in effect used to fund the 'real' price reductions (particularly on power station coal) necessary broadly to maintain sales volume, rather than to increase profitability. Second, this approach implied that the large internal cross-subsidies would continue (as between low-cost opencast sites and deep mines on the one hand, and the bulk of higher cost deep mines on the other). Indeed, the internal cross-subsidy mechanisms were strengthened by the artificial nature of the pricing arrangements under the Understandings. Those required the electricity generators to take the highest-price 'base tonnage' first before the lower-priced tranches became available. Although the prices of power station coal were averaged across all colliery accounts, the gap between the costs of the highest-cost collieries supplying the market and the *marginal* price obtained was very large. As a mechanism for maintaining volume, this was very effective, but it would be difficult to pretend that this was normal commercial behaviour.

Radical Restructuring: Colliery Closures and Manpower Rundown

However, although the Department of Energy grumbled about the extent to which BC used price discounts to stay in the export market and parts of the industrial market, there was no serious attempt by the government to push BC further towards a strategy closer to 'profit-maximisation'. Indeed, there was tacit recognition that the 'de facto' objective pursued by British Coal in this period was by no means a soft option.

The Energy Committee's Report on the Coal Industry (HC165) published in January 1987, made a 'consensus' forecast of demand for BC Coal in 1990 in the range 87–110m. tonnes (HC165 para 57) – in other words, a trend of broad stability, albeit with a degree of uncertainty. Even though in fact total sales volume showed a modest decline, the need to respond to falls in average revenue, while meeting the financial objectives set by government, entailed large increases in labour productivity and reductions in colliery manpower. Further, to accommodate the additional output arising from the completion of the Selby Complex as it reached its planned capacity of at least 10m. tonnes per annum, and the expansion at some of the best reconstructed pits, involved offsetting reductions in higher-cost capacity. For these reasons, the degree of continuing restructuring required was very great. And a great deal was achieved.

By 1989/90, labour productivity was double the level being achieved before the Great Strike; and between the end of the strike and March 1990, the number of operating collieries had been more than halved, and the labour force reduced by over 100,000 men. (See Table 3.8)

Before and immediately after the Great Strike, it was confidently

Table 3.8: Number of Collieries, Manpower and Productivity 1985/6 to 1989/90

	1984/5	1985/6	1986/7	1987/8	1988/9	1989/90
Number of operating collieries (year end)	169	133	110	94	86	73
Colliery manpower ('000) (year end)	171	138	108	89	80	65
Output per man year (tonnes)	*	571	700	789	978	1080

* strike year

Source: *BC Annual Reports*

assumed that, provided the industry's output could be brought into balance with demand, and major inroads made into the closure of uneconomic deep-mined capacity and general over-manning, then indeed British Coal could be turned from 'an institution into a business'. Yet even the huge programme of rationalisation of deep-mined operations proved to be insufficient to reach this goal.

BC's financial results were still far from satisfactory over the four years to 1989/90. (See Table 3.9). There was a huge net loss of £1,899m. before deficit grant. It can be said, by way of mitigation, that this enormous sum can be explained in terms of the exceptional social and other costs arising from the necessary programme of colliery closures, and the excessive levels of interest payments arising from the unrealistic asset base. Moreover, there were substantial operating profits. But this was beside the point. The operating profits came from opencast operations and non-mining activities. As always, the problem was the deep mines, which made an aggregate operating loss over the four years,

Table 3.9:   BC Financial Results 1986/7–1989/90. £ million

	1986/7	1987/8	1988/9	1989/90
Operating Profit/(Loss)				
Deep Mines	41	(67)	132	(149)
Opencast	244	252	272	234
Other activities	32	26	12	(11)
Profit in sales of fixed assets	52	50	82	59
Total	369	261	498	133
Exceptional Items				
Social costs less grants	(197)	(146)	(97)	(257)
Terminal depreciation	(62)	(241)	(172)	(215)
Total	(259)	(387)	(269)	(472) *
Taxation and other items	(12)	(1)	-	-
Result before interest	98	(127)	229	(339)
Interest	(386)	(368)	(432)	(574)
Loss before deficit grant	(288)	(495)	(203)	(913) *

(Derived for Schedule 3 : BC AR 1989/90)

*   The result for 1989/90 quoted above excludes the write down of fixed assets and other provisions embodied in BC's financial reconstruction under the 1990 Coal Industry Act (see below).

with over half the output being unprofitable. Although average colliery operating costs by 1989/90 were 25 per cent lower in 'real' terms than in 1985/6 (a reduction in excess of the government's target), this was matched by the fall in average revenue designed to maintain sales volume, and due in large measure to the price provisions in the Joint Understandings with the CEGB (See Table 3.10). Thus there was little or no progress in improving underlying deep-mine profitability.

Table 3.10: Trends in Average BC Revenue in 'Real' Terms 1985/6 to 1989/90.
(1985/6 = 100)

	Power Station Coal	Average Colliery Revenue
1985/6	100	100
1986/7	93	92
1987/8	86	85
1988/9	83	80
1989/90	77	76

Sources: Power Station Prices: Table 7.2 Coal 'White Paper' 1993 (Cm. 2235) - Average colliery revenue: BC Annual Reports GDP deflator used

Sizewell B and Cheap Imported Coal

The potential threat to BC from imported steam coal as a major issue came into the public domain by an unusual route, namely, the Public Inquiry into the proposal to build a Pressurised Water Reactor (PWR) nuclear station at Sizewell (known as Sizewell B).

For much of the 1980s, it was thought that nuclear power was the main threat to the markets of the UK coal industry. As early as December 1979, the Secretary of State for Energy (David Howell) had announced the aim of starting work on a PWR nuclear station in 1982, and that the CEGB wished to order one PWR a year over the following ten years, a total of 15GW of additional nuclear capacity. In April 1982, the CEGB issued its statement of case and revised detailed design for the first PWR: Sizewell B. The plan was that the ensuing public inquiry would cover the economic and safety case for Sizewell B and the generic issues in such detail that, assuming the inquiry endorsed the case for Sizewell B, the need for such examination for subsequent PWRs would be largely obviated, thereby facilitating and speeding the programme. The government's motivation was political as well as economic. As Nigel Lawson said in his memoirs: 'The PWR was seen as vital to demonstrate to the NUM that coal was not fundamental to the economy any longer, and that nuclear power stations could be built to time and to cost. The need for "diversification" of energy sources, the argument I used to justify the PWR programme, was code for freedom from NUM blackmail' (Lawson p.168). The Sizewell inquiry began in January 1983 and ended in March 1985. The Inspector's Report was not submitted to the Secretary of State until December 1986. The process had been one of almost obsessional thoroughness, and of such duration that it spanned the Great Strike, and the collapse of oil prices. But the government had got the result it wanted: the report concluded that Sizewell B was the lowest-cost option available for meeting generating capacity needs, and that planning permission should be granted. Notwithstanding the fundamental changes that had taken place in UK and international energy markets since the inquiry had began, the Secretary of State (Peter Walker) issued his favourable decision letter almost immediately following receipt of the Inspector's Report, and construction work began in March 1987. But in spite of Mrs Thatcher's continuing enthusiasm for nuclear power, Sizewell B was to prove to be the only nuclear station ordered during her eleven years in office (and the last to be ordered in the twentieth century).

However, before the Sizewell inquiry began, it seemed likely that, if approved, Sizewell B would be followed by further PWR stations. This placed the NCB in something of a dilemma. Not to appear at the inquiry, given the size of the potential threat to UK coal markets, would have seemed a dereliction of duty, yet to have joined the ranks of the opponents would have risked censure from the government. The posture adopted by the NCB was therefore that, given the centrality of coal prices in the economic case for nuclear stations, the Board should offer its assistance to the inquiry on this important matter. This also appeared to be consistent with the NCB's interest, since its views on future coal prices were lower than those held at the time by the CEGB and the Department of Energy, and would, therefore, other things being equal, weaken the case for Sizewell B.

This episode serves to illustrate, in an emphatic way, the extent to which views changed on the future prices of internationally traded steam coal. The NCB case to the Sizewell B inquiry, presented by the author (Proof of Evidence: Coal Price Prospects and Availability of Coal in the UK Power Generation Market: M J Parker; February 1983: Sizewell B Public Inquiry) said in summary that

> the NCB do not see the availability or cost of NCB coal as being relevant to the CEGB's economic case for Sizewell 'B' but they agree that the

future international price of coal (which will determine the value of coal in the UK) is a central issue. The future trends of international coal prices are very uncertain and subject to wide margins of error. However, there are now generally lower expectations of the rate of increase in the demand for coal in Western Europe. Further, additional supplies of relatively low-cost traded coal can be brought into Western Europe on a shorter timescale than additional demand for some time ahead. These factors are likely to lead to relatively low escalations in the underlying trend of the prices of international coal delivered into Western Europe. Therefore the NCB would not regard as robust investments (whether in coal production or utilisation, or in alternative fuels) which depended for their viability on an early or rapid escalation in international coal prices, or upon this escalation continuing indefinitely into the future.

The Inspector (Sir Frank Layfield) accepted that the value of coal in the UK would reflect international prices, rather than NCB costs; and adopted NCB methodology on future prices for international coal. The estimates for 'ARA' prices (that is, prices of coal delivered to Amsterdam, Rotterdam, Antwerp) in March 1982 $ are shown in Table 3.11.

Table 3.11: 'ARA' Steam Coal Price Estimates: Sizewell 'B' Inquiry. March '82 $/tce*

	1990	2000
Dept. of Energy (Mid-point of range)	92	139
CEGB (Central case)	82	105
NCB (Central case)	60	86

* tce = theoretical standard coal of 29.3GJ/tonne

Source: Sizewell B Public Inquiry: Inspector's Report. Chapter 73: Annex 73.1

These prices compared with actual prices at the time (1981 and 1982) of around $70/tce. Subsequent events were to show all these projections (including the NCB's) to be absurdly high. The collapse in oil prices in 1986 lowered the price of fuel oil, which set a new ceiling for steam coal prices; and, more fundamentally, over time it became clearer that internationally traded steam coal was a commodity market with an inherent tendency to over-supply as a result of coal-on-coal competition and the continuing availability of intra-marginal supplies. Between 1982 and 1990, the $ ARA price fell by 42 per cent in 'real' terms, and the ARA price denominated in sterling fell in 'real terms' by 59 per cent. (See Table 3.12). The economics of the UK coal industry were thereby utterly changed.

Table 3.12: Contract (c.i.f.) Price of Imported Coal for Power Stations in EU

	1998 $/TCE	*1998 £/GJ*
1982	111.9	2.97
1983	89.1	2.53
1984	75.8	2.51
1985	74.3	2.14
1986	67.7	1.93
1987	58.6	1.31
1988	60.8	1.34
1989	63.0	1.52
1990	65.5	1.22

(Derived from IEA data: *Coal Information*, using US and UK GDP deflators. Prices based on tce = 29.3 GJ/tonne)

British Coal was protected from the full impact of this trend by the price mechanism in the various Joint Understandings. Over the four years 1986/7 to 1989/90, even after allowing for BC's lower internal total transport costs, BC's prices to power stations were some 40 per cent higher than those of imported coal. Although the total displacement of BC coal by imports was not a practicable option, nevertheless, in broad terms, imports set the real value of UK coal, which by the end of the 1980s was no more than half the level obtained before the Great Strike. The great scale of this change took some time to be generally accepted. We have already noted that, over the period 1985/6 to 1989/90, the average cost of BC's deep mines was reduced by 25 per cent in real terms, reflecting the massive restructuring which had taken place; but also that average revenue per tonne was reduced by a similar amount as part of the policy to contain reductions in sales volume. But, the 'commercial' value of UK coal, as measured by the price of imported coal, was falling even faster. By the end of the 1980s, in spite of the enormous efforts involved in doubling productivity, closing over half the operating collieries and reducing the labour force by over 100,000, BC was no nearer fully viable operation than it was before the Great Strike.

Capital Investment Falls while Productivity Rises

A further symptom of this deterioration in the economic fundamentals was the increasingly unfavourable outlook for investment, which declined over the period. (See Table 3.13)

Table 3.13: Capital Expenditure on Mining. £million

	Major Colliery Projects	Other	Total	Index 'Real Terms'* 1985/6 = 100
1985/6	236	409	645	100
1986/7	280	363	643	97
1987/8	318	322	640	92
1988/9	249	298	547	73
1989/90	177	299	476	60

*   Using GDP Deflator

Source: *BC Annual Reports*

By 1989/90, total capital expenditure on mining (almost entirely on deep mines) had fallen by 40 per cent in 'real' terms compared with 1985/6 (and 60 per cent compared with 1980/1). New approvals of major projects fell rapidly, although the total expenditure was sustained for several years by the large sums being spent on the Selby project, and which was effectively fully committed by 1985/6. Nevertheless, in due course, total expenditure began to reflect increasingly stringent procedures adopted by BC. To a considerable extent, these procedures were based on the 'cost criteria' originally set out in the 'New Strategy for Coal', formulated in 1985 after the end of the Great Strike, and to which reference has already been made. Although some changes were made from time to time, the main features of this system were an upper limit of cost of £1.50/GJ at collieries which were to attract major investment, and a cost limit of £1/GJ for any 'incremental' output arising from major projects. As these criteria were not indexed to general inflation, they were tightened in 'real terms' over the period. The £1.50/GJ upper limit remained well below *average* colliery revenue.

The cost criteria also had the effect of assisting the ordering of priorities for deep-mine investment, as set out in the 'New Strategy for Coal', namely.

Firstly, expenditure to reduce costs and increase productivity at those existing mines able to operate at competitive costs, and to sustain ongoing operations at such collieries;

Secondly, major projects to provide new and replacement capacity at the best of existing mines when the marginal cost of such capacity is low;

Thirdly, new mines capable of providing competitive output in the longer term. (Quoted in MMC Cm. 550. para. 3.29)

These priorities for investment were also guided by a view that capacity planning should be influenced more by considerations of profitability than by output targets. As BC said to the MMC in 1988,

> There is ... a wide range of possible outcomes in the size of BC's own market, and it would thus, BC believes, be quite impracticable to fix upon a particular level of future demand for the purpose of planning capacity and investment. BC holds, furthermore, that such an approach, even if practicable, would be contrary to its objectives, which are clearly designed to put 'money before tonnes'. BC therefore rejects the notion of planning investment to produce a given target output. As a consequence, it also rejects 'gap-ology' (i.e. the determination of a shortfall between projected supply and projected demand) as a procedure for determining 'capacity-related' investment. (MMC Cm. 550.para. 3.16)

This clear statement of investment policy principles (with which the author was associated), had an ambiguous relationship with the overall 'de facto' management of the industry which, as we have noted, was to maintain the deep-mined industry at the highest practicable level consistent with a broad supply/demand balance and an overall financial breakeven. Schemes were often advocated by operating areas to Headquarters on the primary grounds that they were required to maintain the output of the Area, or that they were 'needed' to meet demand. The internal 'cultural revolution' required to turn BC from 'an institution into a business' still had some way to go. Nevertheless, trends in capital expenditure in five years after the Great Strike not only showed the 40 per cent 'real' fall noted above, but also the changes in priorities of the 'New Strategy for Coal'.

A particular emphasis was placed on 'Heavy Duty' face equipment; a term applied to coalface installations with powered roof supports, conveyors and power loaders of a type larger, more powerful and more robust than those traditionally used. At the same time, there was a move away from advance longwall to retreat longwall mining. On an advancing face, the extraction of coal is carried out simultaneously with the driving of the access roads at right-angles to the face line; whereas, on a retreating face, the extraction of a panel of coal takes place using pre-formed roadways driven to the ultimate boundary, from which coal-working commences. Provided that high rates of roadway drivage can be achieved, a retreating system allows significantly higher levels of face outputs per day.

Between 1982/83 (the last complete financial year before the Great Strike), the proportion of Heavy Duty faces increased from 6 to 48 per cent, and retreat faces from 21 to 37 per cent of the total. (See Table 3.14)

These developments, which flowed partly from investment (both in

Table 3.14: Coal Face Productivity to 1987/8

	1982/3	1985/6	1986/7	1987/8
Number at year end				
Heavy duty faces	32	87	119	117
'Conventional' faces	542	294	186	129
Total	574	381	305	246
Heavy duty as % of total	6%	23%	39%	48%
Advancing faces	454	278	202	155
Retreating faces	120	103	103	91
Total	574	381	305	246
Retreat as % of total	21%	27%	34%	37%
Daily output per face (tonnes)				
Heavy duty	1390	1393	1459	1496
'Conventional'	711	766	900	1003
Advancing	710	809	984	1106
Retreating	872	1042	1260	1385
Overall Average	730	869	1067	1205

Source: MMC Cm. 550 Tables 4.8 and 4.9

colliery infrastructure under 'Plan for Coal', and in new coalface machinery) and partly from colliery closures, were largely instrumental in the very large increase in overall output per man year (see Table 3.8), since the large increase in the output per face enabled the optimum level of output at any continuing colliery to be achieved with much lower supporting manpower 'elsewhere below ground'. This whole process was also assisted by the operation of incentive wages schemes, which enabled real wages to rise substantially, while at the same time reducing wages costs per tonne.

BC also promoted other ways of increasing productivity. For example, in 1986 BC's Nottinghamshire Area Director (Albert Wheeler, later a BC Deputy Chairman) put forward proposals which included six-day working, extending coal winding time per day, greater use of outside contractors, the introduction of new incentive payment schemes for mineworkers, and increased use of flexible production techniques such as roof bolting and trackless underground vehicles. In evidence to the Energy Committee, Sir Robert Haslam expressed broad agreement with these proposals (HC 165 para. 79). Nevertheless, the episode of what became known as the 'Wheeler Plan' illustrates the continuing sensitivity of industrial relations issues. Much of the thinking about 'flexible working practices' was anathema to the NUM and their allies in NACODS, although the UDM remained more accommodating on these issues. Consequently, BC's public position was that the 'Wheeler Plan', although containing many useful ideas, was not official BC policy; and BC continued over the following few years to adopt a cautious pace on the introduction of new mining and working practices at collieries.

The MMC presented its report on the investment programme of British Coal (Cm.550) in January 1989. Its conclusions in the area of strategy (that is conclusions other than those concerned with improvements to internal procedures) appeared contradictory. While it encouraged BC to 'think more imaginatively about the ways in which things can go wrong' (para 3.41), it also seemed to suggest that BC were too risk-averse: 'BC's stated policy of "minimum regret" for investment decisions properly, in our view, concentrates short-term attention on cost reduction at existing mines. But such a concentration … could lead to the avoidance of bold decisions, the lack of which might be regretted in the longer-term future' (para 3.40). This appears to suggest that BC should have looked more favourably at new mine prospects. Yet the detailed information considered by the MMC scarcely encouraged such a view. The economic case for the huge investment in the Selby new mine complex, which had been the showpiece of the 'Plan for Coal', had depended to a considerable extent upon the assumption (widely held in the 1970s) that oil and coal prices would continue to rise in real terms – a view no longer tenable by the mid-1980s (by which time the investment in Selby had been largely committed).

In this context, the case of the Asfordby New Mine was symbolic. This was the surviving mine in the originally proposed North East Leicestershire Prospect (generally known as the 'Vale of Belvoir' development) which had earlier been rejected by the government on planning grounds. As noted in Chapter 2, the government invited BC to resubmit a somewhat modified scheme for Asfordby which was approved in 1985 after the end of the Great Strike. Indeed, the BC Chairman (Ian MacGregor) and the government both presented this as a 'reward' for those miners in Nottinghamshire and Leicestershire who represented the core membership of the new Union of Democratic Mineworkers (UDM), and who had continued to work during the strike. From the outset, there were doubts about the quality of the coal and the financial return. These doubts were overriden by BC's Capital Investment Committee in November 1985, given that the 'Office of the Chief Executive' (namely Ian MacGregor) had already signified approval, and 'bearing in mind the consequences for the industry as a whole'. This was code for support for the working miners and UDM: a policy strongly advocated by Mrs Thatcher at the time. The Asfordby new mine project continued to be as much a matter of industrial politics as of commercial judgement. With more realistic views on coal prices, the annual review of the project in 1987 showed a DCF return after risk assessment of 6.4 per cent (against the BC requirement of at least 10 per cent return) even when expenditure already incurred or committed was disregarded, and in spite of an increase in the planned output from 2.2 to 3.8m. tonnes per annum. In 1988, the author's Economics Unit questioned whether a satisfactory return on investment could be achieved. Nevertheless, the project was allowed to continue on the grounds that to discontinue it would cause more problems than going on. Even on capital investment, there were limits as to how far BC could behave like a commercial business (See MMC Cm. 550 Appendix 5.2).

During the period 1985–90, there were various other proposals for new mine development, which came to nothing. Most of the pressure for these projects came from senior management in the coalfields (and their supporters at BC headquarters) who wished to establish both for themselves and their workforce, that their areas had a long-term future. In 1987, BC gave conditional approval for a new mine at Margam in South Wales, to provide 1.2m. tonnes per annum of prime coking coal, but in spite of the high quality of the coal, and the favourable productivity projected, there was substantial price risk, so that BC approach was made conditional on receipt of government grants as if BC qualified for development area finance (which had not been allowed hitherto). In the event, as grant aid was not forthcoming, and as British Steel were unwilling to enter into a satisfactory long-term contract the project did not proceed (see MMC Cm. 550 Appendix 5.6). In 1987, BC submitted a planning application for a new mine at Hawkhurst Moor, designed to work the Warwickshire Thick Coal next to Daw Mill colliery in Warwickshire, which was a UDM pit already subject to a major project for expansion of output. Hawkhurst Moor was the subject of a public inquiry in 1989, but the Inspector's report rejected the application on planning grounds, and the project was not taken further. The reasons why BC proceeded with their case at the public inquiry at that time is not clear (although 'being seen to support the UDM' was involved). The author took the unusual step of briefing

BC's Counsel at the inquiry on the dubious commercial economics of the project, and the unlikelihood of its proceeding. Finally, in the late 1980s, following an exploration programme, the Nottinghamshire management put forward proposals for a new mine at Witham near Newark, but this was not taken further, on the grounds of speculative economics.

Asfordby was the last new mine to be constructed by BC, and on strictly commercial grounds even this single new mine project should have been discontinued (Asfordby was subsequently closed after privatisation, in 1997). Although opinion within BC was not unanimous on the issue, by 1990 it was becoming increasingly clear that further new mines were unlikely to be sunk for the foreseeable future. For an extractive industry, and with a high proportion of old collieries, this was a tacit admission that the economic fundamentals pointed to inevitable long-term decline (although at that time there still appeared to be a number of opportunities for productive investment at the best of existing collieries).

Government Financial Support

The government welcomed the downward trend in investment by BC, particularly in major new schemes on primary capacity. Nevertheless, this was not reflected in any reduction in BC's External Financing Limits (EFLs). Indeed, large upward adjustments were made each year to the External Financing Limit (EFL) initially set. (See Table 3.15)

The apparent willingness of the government to accommodate year-by-year these very large additional requirements for government funds reflects, above all, a willingness to finance the restructuring process

Table 3.15: BC's External Financing Limit 1986/7 to 1989/90. £million

	Original EFL	Adjusted EFL	Out-turn EFR*
1986/7	730	825	902
1987/8	727	920	
1988/9	670	850	
1989/90	720	1143	1292

EFL = External Financing Limit
EFR = External Financing Requirement

* Where EFR exceeded (adjusted) EFL

Source: *BC Annual Reports*

and the redundancy and other costs associated with the very large reduction in colliery manpower during this period. This was acknowledged in successive BC Annual Reports. In 1986/7, the additional finance requirement arose from the coal price reductions required following the collapse of oil prices (see above) and 'the heavy social costs associated with restructuring the industry'; in 1987/8, additional needs arose from 'the continuing weakness in the market, coupled with higher than anticipated costs associated with restructuring the industry' (with the labour force reduced by 21,400 against the original expectation of only 7,000); the report for 1988/9 refers to the way in which difficult market conditions compelled the Corporation to accelerate the restructuring of its operations; and in 1989/90, BC reported that it had continued the programme of restructuring its operations thereby incurring higher than planned costs.

This emphasis on government funding of restructuring can be seen also in the make up of government grants over the five years following the strike. (See Table 3.16)

Table 3.16: Government Grants to BC 1985/6 to 1989/90. £ million

	1985/6	1986/7	1987/8	1988/9	1989/90
Social costs	450	531	435	255	619
Contribution to increased pensions	63	63	42	42	-
Deficit Grant	50	288	200	-	*
Total	563	882	677	297	619
RMPS	566	611	290	169	129

* Figures for 1989/90 exclude special arrangements arising from the Financial Reconstruction in the Coal Industry Act 1990

Source: *BC Annual Reports*

Grants for Social Costs arising from the provisions of successive Coal Industry Acts included expenditure incurred in connection with the redundancy, redeployment or early retirement of employees and (from 1987/8) the cost of promoting alternative employment in mining areas.

In addition, payments under the Redundant Mineworkers' Payments Scheme (RMPS), although administered by BC, were paid directly by government (RMPS was closed at the end of 1986/7, but payments for previous redundancies continued). These payments for manpower reductions were very large, amounting to over £4 billion ('money-of-the-day') over the five years after the strike, and representing 85 per cent of the total revenue support given to BC (before taking account of the effects of the financial reconstruction in the Coal Industry Act 1990). Over the five-year period colliery manpower was reduced by 106,000, so that the above expenditure was equivalent to £60,000 per job lost (in 1998 money values). (See Table 3.17)

Table 3.17: Government Financing of Manpower Rundown. Total of Social Cost Grants and RMPS Payments. £ million

	'Money of day'	*1998 Money Values*
1985/6	1016	1703
1986/7	1142	1859
1987/8	725	1121
1988/9	424	615
1989/90	748	1015
Total	4055	6313

The final element in government financial support in this period was the financial reconstruction embodied in the Coal Industry Act 1990 (which received Royal Assent in March 1990), under which the Secretary of State for Energy was empowered to make a Deficiency Grant to BC not exceeding the accumulated deficit at 31 March 1990 to cover

- a write down of £2,606m. of the Corporation's fixed assets, reflecting the dramatically lower world coal price
- a provision of £1,976m. for potential liability for industrial deafness claims and concessionary fuel arising from past service
- other losses up to March 1990 of £1,571m. arising from high interest and restructuring costs.

In addition £471m. of grant for concessionary fuel or cash alternative would be payable under the 1977 Coal Industry Act.

The Coal Industry Act 1990 also extended to March 1993 the time limit for the payment of restructuring grants under the Coal Industry Act 1987, and raised the limit for the aggregate amount of grants under that Act. But no further grants would be available to cover operating losses incurred by BC after 31 March 1990.

The effect of the 1990 Act was to 'wipe the slate clean' so far as past losses were concerned, and to reconstruct the Corporation's balance sheet by eliminating borrowings related to past losses and over-valued assets. This was a sensible measure, since BC's interest burden had become disproportionate in the light of the industry's economic prospects. In practical terms, there was a large reduction in capital charges; namely in depreciation and interest, the total of which fell by over £600m. as between 1989/90 and 1990/1. This was an ongoing benefit to BC, for future years, and was certainly an essential pre-condition to the establishment of profitable operations in the difficult transitional period between ESI privatisation in 1989/90 and BC's own privatisation in 1994 (as we shall see later).

But there was a price to be paid. The write down of assets represented almost two-thirds of the previous asset base. This was an admission that most of the expenditure incurred under 'Plan for Coal' could not provide a commercial term, in the new market conditions following the collapse of oil prices in 1986, and the continuing fall in international coal prices. By the same token, the case for investment in new and replacement capacity was further dramatically weakened, with all the implications for long-term decline. The revaluation of assets also drew attention to the extent to which even existing capacity was still uneconomic. The government had insisted that BC should value each colliery on the basis of its individual projected cash flow. Barely 20 out of 75 collieries were able to show a positive present value on this basis – a clear indication of how much further the rationalisation process would have to go before BC could be said to be operating as a commercial business.

Environmental Concerns about Coal

In addition to growing evidence of the continuing underlying economic weakness of BC's deep-mined operations, there was another unwelcome trend. In ways which were to have great long-term significance, the 1980s saw a steady and significant increase in environmental concerns relating to coal, and a change in their character. The 1981 report of the Commission on Energy and the Environment on the environmental impacts of coal (the 'Flowers Report') dealt almost exclusively with problems related to coal production: tipping, opencast operations, the local impact of new mines, and subsidence. Questions of atmospheric pollution and wider environmental concerns received little discussion. The Commission considered that 'there is no longer major cause for concern on health grounds over ambient levels of smoke and sulphur dioxide from coal combustion', and 'the situation is also satisfactory as regards corps and vegetation'. They also concluded that 'nitrogen oxides from coal combustion have not been identified as the source of any pollution problems in the UK', and that 'no deterioration of the climate due to the build up of carbon dioxide in the atmosphere has yet been established' (Commission on Energy and the Environment 1981, paras 22.111, 22.113 and 22.117). By the end of the 1980s the focus of environmental policy had greatly changed, and to the disadvantage of UK coal.

The first major issue to arise was 'acid rain' and SO_2 emissions. When examined by the Environment Select Committee in 1984, the issue was seen primarily in terms of 'establishing the facts' through research. However, due mainly to political pressure in Germany, which took massively expensive action to fit fluegas desulphurisation (FGD) equipment at virtually all coal-fired power stations ostensibly to 'save the German forests' but in a way compatible with protection of the German coal industry, there was increasing pressure for other governments, particularly the UK government, to come into line. The Community's Large Combustion Plant Directive (LCPD), agreed in 1988, committed the UK to progressive reductions in SO_2 emissions at power stations: namely, reductions, compared with a 1980 base, of 11 per cent by 1993, 40 per cent by 1998 and 60 per cent by 2003. At the time it was generally assumed that the position of UK coal would be largely safeguarded by a programme of 12GW of FGD plant. However, as electricity privatisation approached, it became clear that this programme was not a firm commitment. More potentially serious for the longer term was the rising interest in the 'Greenhouse Effect' and the role of CO_2 emissions in leading to global warming. The Energy Committee produced a report on the 'Energy Policy Implications of the Greenhouse Effect' (HC 192, July 1989). Neither the Committee nor the government were willing at that time publicly to draw severe conclusions for coal. As the Committee said, 'we agree with the Secretary of State that the country cannot afford to turn its back on its largest indigenous source of fuel resources' (para 93). Nevertheless, this was the beginning of a major concern for coal, which, by reason of its fundamental chemical composition, was bound to emit more CO_2 than competing fuels for any given level of end-use efficiency.

Government Policy after the Great Strike

Over the five years following the Great Strike, government policy was aimed at moving British Coal towards viability, primarily by facilitating massive restructuring and the associated manpower rundown: albeit in a controlled way. There was no longer any suggestion of general government support for UK deep-mined coal on 'energy policy' grounds. Objectives given to successive BC Chairmen placed increasing emphasis on financial results; and downward pressure on the industry's capital expenditure in new and replacement capacity was reinforced by the capital reconstruction in 1989/90. Although BC benefited from artificially high prices for power station coal, there was effectively no support for ongoing operations by *direct* subsidy: the very large level of government finance made available to BC was concentrated on funding generous redundancy payments and other costs arising from manpower rundown and colliery closures. Moreover, government grants and other support arrangements were so organised that BC management was not inhibited in any way from restructuring by the cost of doing so. External Financing Limits were altered as required.

Although the result of these policies was a rundown of over 100,000 in miners' jobs over the five years from March 1985 to March 1990, there appears to have been no attempt by government to weigh the costs of restructuring and redundancy against wider social and unemployment costs. Indeed the rapid rundown of coal industry manpower was organised in a way which effectively precluded an overall calculus. In its evidence to the Energy Committee in 1986/7, the Department of Energy made it clear that although 'decisions on individual pits and the phasing of closures are the responsibility of the NCB', 'the problems facing a local community as a result of pit closure need to be faced and solved directly; for the country as a whole it is no solution simply to keep open pits which cannot produce coal economically' (HC165, Evidence paras 5.1 and 5.3). The Committee urged that if the macroeconomic disbenefits of closure outweighed the benefits, it was the job of government to right the balance (HC165, Report, para 131). In its formal response, the government avoided the question, by merely stating that it could not be the responsibility of British Coal to make such judgements. (HC 387 para. 5.3)

As we have seen, the public subsidies to facilitate job losses in mining were very large. In effect, the government had turned on its head the arguments used in the 1960s and 1970s to mitigate job losses by attaching a 'shadow value' to labour in areas of high unemployment which was below accounting costs. By contrast, the implication of the government's policy in the 1980s (and early 1990s) was that there was overriding public benefit in reducing employment in coal mining, such as to justify public expenditure equivalent to around £60,000 per job lost (1998 money values).

Yet this apparently ruthless policy towards manpower rundown had been mitigated by the government's caution in a number of respects, designed to make the process more politically acceptable and to prevent any resurgence of serious industrial disputes, however improbable these might appear given the scale of the NUM's defeat in the Great Strike.

First, the redundancy terms for mineworkers were kept as generous as necessary to ensure that the men themselves would vote for closure, thereby allowing a policy of voluntary redundancy to be sustained in a way which made Union opposition to closures ineffective, and minimised the possibility of serious disputes over manpower rundown. The Redundant Mineworkers Payment Scheme (RMPS), which had been in operation for nearly twenty years, and had been directly funded by government, came to an end in March 1987, and was followed by four 'British Coal' schemes, which were still effectively funded by government, but indirectly as part of the restructuring grants made available to BC. Between April 1987 and August 1989, the average age of redundant mine workers was 41 years, and the average lump sum payment was £18,000 (money of day). In addition to the generosity of the payments, a further incentive for men to take redundancy arose from the policy of giving each successive scheme a terminal date, with no assurance that a further scheme would be introduced.

Second, the increasing use of incentive payments associated with high levels of productivity meant that those mineworkers remaining enjoyed high levels of wages. There was no suggestion that BC should seek to reduce its costs by restricting earnings.

Third, the government was insistent that BC should continue to process closures through the Modified Colliery Review Procedure (MCRP), which had been in place since the end of the Great Strike. This was in spite of the increasing difficulties experienced by BC in reconciling the length of this procedure (which involved nine stages lasting up to nine months after a closure had been recommended to the Corporation) with the timing of redundancies agreed at local level. Indeed, when in April 1990 Sir Robert Haslam made personal representations to Mrs Thatcher on this matter, he was told in no uncertain terms that the government would not change its position.

Fourth, the government continued to support the activities of British Coal Enterprise (BCE), which had been set up in 1985, with the support of Ian MacGregor, to encourage new employment opportunities in mining areas, and to help settle former mineworkers into new jobs. By March 1990 BC claimed that the eventual number of new employment opportunities resulting from BCE's activities had reached nearly 71,000, including 39,000 in small businesses which had received £69m. of loans or equity from BCE. Although it is not possible to obtain a precise estimate of how much of this activity would have occurred without BCE's help, its activities did do something to soften the impact of colliery closures, and to make BC's manpower rundown less politically contentious. (*BC Annual Report 1990/1*, p.29)

A particular aspect of the government's caution towards industrial relations in the 1985–90 period concerned its protective attitude towards the UDM. This was very much the result of the personal views of Mrs Thatcher. As she was to write:

> The memories of the year-long strike were unforgettably etched on my mind. I kept in touch with Roy Lynk, the Nottinghamshire leader of the UDM, and I made sure that both Cecil (Parkinson) and John (Wakeham) understood my feelings about the need to protect the interest of his members. First, I felt a strong sense of obligation and loyalty to the Nottinghamshire miners who had stayed at work in spite of all the violence the militants threw at them. And, second, I also knew we might have to face another strike. Where would we be if we had closed the pits at which moderate miners would have gone on working, and kept more profitable but more left wing pits open? (Thatcher p.686)

Thus, Mrs Thatcher's wish to protect the UDM pits was, in part at least, due to a continuing anxiety that a further coal strike might occur if BC's approach to closures was too insensitive. The government also sought to ensure that power station coal stocks were rapidly rebuilt after the end of the Great Strike, and maintained at a high level thereafter. In the five years after the strike, financial year-end coal stocks held by consumers (nearly all by power stations) averaged 25m. tonnes – significantly higher than the average for the five years before the strike. As important, the government consistently used its influence with the nationalised electricity generators, particularly in the period immediately following the collapse in international oil and coal prices in 1986, to ensure that the price and volume of UK coal supplied to power stations were sustained at substantially higher levels than would have obtained in a 'free market'.

Continuing Weakness of BC's Position

Assisted by the caution and financial generosity with which government policy had been conducted, by the end of the 1980s BC had carried through an enormous restructuring programme, which had lead to a doubling in productivity and large reductions in operating costs in 'real' terms. But BC had not yet become a fully commercial business. The internal culture still tended to put 'tonnes before money': that is financial considerations were still seen as constraints rather than overriding objectives. The case for major investment in new and replacement capacity to sustain long-term output had become untenable, and the value of BC's output was declining with the falling price of coal imports. BC was still 80 per cent dependent upon the 'special relationship' with the electricity supply industry, which sustained BC sales volume and prices at a significantly higher level than fully commercial arrangements would have given.

Further, public policy concerns which had previously supported the industry's case for protection carried much less political influence, as a result both of events and of the policies of the Thatcher governments. No longer was a special value placed on the coal industry as a means of providing employment in those regions which were economically disadvantaged; the defeat of the Great Strike in 1984/5 reduced the fear successive governments had had of the power of the NUM to disrupt electricity supplies; the collapse of oil prices in 1986 and associated reductions in internationally-traded coal prices had not only increased market pressures on BC, but also had in large measure made UK and other Western governments much less interested in 'security of supply', thereby downgrading the formerly perceived 'strategic' value of UK coal; and UK coal was increasingly seen as environmentally undesirable.

In this state of fundamental economic and political weakness, and with its transition to a commercial business still far from complete, the UK coal industry faced the next great challenge: electricity privatisation.

CHAPTER 4
ELECTRICITY PRIVATISATION AND THE 'COAL CRISIS' OF OCTOBER 1992

Emerging Threats to British Coal

The privatisation of the electricity supply industry (ESI) was to prove to be adversely momentous for the future of the UK coal industry. The process began in 1987, when the Conservative Party manifesto for the General Election (which again resulted in a Conservative administration) included a pledge to privatise the electricity industry. In February 1988, the government published two White Papers (Privatising Electricity: the Government's Proposals for the Privatisation of the Electricity Supply Industry in England and Wales (Cm. 322) and Privatisation of the Scottish Electricity Industry (Cm. 327)) which set out details of the proposed structure which would operate from Vesting Day: 1 April 1990. For England and Wales, the twelve Area Electricity Boards became Regional Electricity Companies (RECs), and retained the monopoly local distribution systems. While large consumers with demand greater than 1 MW would be able to choose their supplier from Vesting Date, RECs would continue to have a monopoly franchise for demand up to 1MW, but reducing to 100 kW in April 1994, and planned to be removed completely in 1998. Whereas the RECs were to be sold intact, the CEGB was to be divided into three parts: the National Grid Company (NGC) was separated from generation, where power stations were placed in two generating companies, National Power, the larger of the two (which would include all the nuclear power stations), and PowerGen. An Electricity Pool was to be set up to act as a market clearing mechanism to deal with dispatch and the determination of wholesale electricity prices. During 1989, the nuclear stations were withdrawn from the privatisation process, which otherwise underwent little fundamental change from the proposals in the White Papers. The launch of the new organisations was achieved in April 1990 as planned, with flotation following.

During this three year period of 1987–90, decisions relating to electricity privatisation were taken which were also to have a profound effect on the UK coal industry. For some time the perception at British Coal of the likely dangers was very mixed. The Energy

Committee announced an enquiry in December 1987 (that is before the government's White Papers were issued in February 1988) into 'The Structure, Regulation and Economic Consequences of Electricity Supply in the Private Sector' (Report HC 307 published July 1988), and BC submitted a Memorandum to the Committee in January 1988 (also appears at Appendix 3.1 to MMC Cm. 550), in which the lead was taken by BC's commercial and economic functions. The gist of the argument put forward was that there should be continuity of the relationship between the UK coal industry and the ESI, to the mutual benefit of both industries. As BC's memorandum stated:

> British Coal's present concern is not with the change of [ESI] ownership as such but that it raises serious questions about the conditions of future coal supplies. The form of British Coal's commercial arrangements with the Generating Boards has reflected the assumption that both parties would remain in common public ownership. These arrangements are not therefore in the form of binding contracts, but mutually-agreed guidelines as to how the two organisations do business together, and as such they have worked well, but the change of ownership now proposed for the ESI requires the negotiation of a new and binding coal supply contract. ... We argue that a long-term contractual relationship between British Coal and British electricity should be settled as soon as possible, and well before any formal change of ownership, so that investors as well as suppliers know exactly where they stand on an issue so vital to both (para 1).

This document was premature, in that it was written before the planned structure of the privatised ESI was known – indeed, the assumption seemed to be that BC would continue to negotiate with a monolithic (but privatised) CEGB. There were two other assumptions made by BC at the time, which in future years were shown to be invalid. First, in 1988 BC assumed that the ESI's high dependence on coal would continue into the 1990s, given the uncertain future of nuclear power, the unlikelihood of any return to fuel oil or the adoption of the use of natural gas, then generally regarded as a 'noble' fuel to be reserved for domestic and other 'premium' uses: whereas, as we shall see, ESI privatisation was to be followed by a large move away from coal into gas-fired generation. Second, BC saw the main threat as additional imports of steam coal which was then being traded internationally at (in BC's view) unsustainably low prices, which, together with the $/£ exchange rate in 1987 and 1988 made the then sterling price of imports unrepresentative of future levels: BC's 'central' estimate of future price levels was some £1.40/GJ in 1987 money values – a level which, in real terms, proved to be over *double* the actual price levels which obtained in the late 1990s.

At the time, BC's view of the future market appeared reasonable. The Energy Committee endorsed its view on the continuing interdependence of the ESI and the coal industry (HC 307 para 129), and noted that there was a widespread expectation both within and outside BC that future international coal prices would rise (para. 132). However, the Committee noted the problems that were involved in moving from the 'Joint Understandings' to commercially-binding contracts, and that the privatised generators would 'in the absence of the political pressures that helped to forge the Understanding in the first place ... be looking for contracts that more closely reflect the present and foreseeable international market for coal' (para 131). Moreover, the new generators might not be willing to have such a high proportion of their coal supply tied to medium- to long-term contracts with BC (para 134). While the Committee agreed on the desirability of an early start to contract negotiations between BC and the to-be-privatised generators (para 136), it found it difficult

> to avoid the conclusion that, under the competitive bargaining arrangements that will come with the process of electricity privatisation, a smaller tonnage of British coal will be sold under term contracts to the private generating companies than is currently being supplied under the Joint Understanding and the pre-1988 arrangements with the SSEB. The BCC's *guaranteed* market will therefore be smaller than it is today. It also seems that the average ex-mine price likely to be secured by the Corporation in its term contract sales will be lower than that paid by the ESI today (para 139).

Clearly, this led to dangers (as the Committee saw it) of significant deep-mine closures, and prompted the Committee to recommend that the government took a view on the 'strategic role for the British coal industry in the 1990s'. Contracts between BC and the to-be-privatised generators 'could implicitly express a view on the strategic role of indigenous coal in the early and middle 1990s' (para 159).

In 1988 the government was in no mood to enter into long-term commitments to protect the coal industry on 'strategic' grounds. The serious business concerned the future BC/ESI contracts which would have to replace the 'Understanding'. For its part, BC did not show a singleness of purpose on this crucial issue at this critical time. There were a number of reasons for this. First, the preoccupation of the numerically dominant senior mining engineers, both on the Board and in the Coalfield Areas, was with carrying out the massive restructuring programme of closures, organising manpower losses within a framework of 'voluntary' redundancy, and achieving productivity improvements and operating cost reductions. Secondly, as we have already noted, the internal cultural revolution, designed to move management's priorities away from 'tonnes' to 'money' was very incomplete, so that there was no clear internal agreement about whether the priority in the negotiations was to secure volume and maintain market share, or to concentrate on the most profitable sectors of the market supplied by the most profitable collieries. Thirdly, there was a general lack of internal understanding on the emerging imbalance of market power as between BC and the to-be-privatised generators. This issue became clearer as ESI privatisation approached. Nevertheless this lack of a clear view was damaging. Malcolm Edwards (the Commercial Director) and the author became increasingly convinced that there was going to be a great imbalance of power against BC and in favour of the generators, since alternative supplies for the generators were far more abundant than alternative markets for BC. However, BC's Chairman, Sir Robert Haslam, appeared to take the view that the generators 'need us as much as we need them'. Fourthly, it became evident that the government placed high priority on the success of the forthcoming ESI privatisation and would look with disfavour on any attempt by another nationalised industry to make that task more difficult. BC was too dependent on government favour for this to be ignored.

Early in 1989, Malcolm Edwards wrote an internal paper on the approach that should be adopted to the new ESI contracts. This emphasised the need to avoid 'fragmented' arrangements (eg individual pit-to-power station deals) which would have enabled the generators to extract most of the financial surplus. The paper also stated that the then current volumes and prices provided by the 'Joint Understanding' (i.e.75m. tonnes/year at an average pithead price of £1.72/GJ at 1988 price levels) were not sustainable. Average prices of £1.50/GJ (1988 money values) were needed to ensure that collieries *on average* would operate profitably. The paper advocated that BC should seek publicly to advance the case for BC's 'unique sales proposition' in terms of security of supply, and predictability of price (including the avoidance of exchange rate risk) as a means of providing stability to the new privatised electricity system. Attempts to 'sell' this approach met with mixed reactions. In January 1989 Malcolm Edwards and the author made a presentation at a meeting of the Chairmen of the Area Electricity Boards (soon to become the RECs), with a view to pointing out the potential benefits of price stability for domestic electricity consumers. This presentation was sufficiently well received to encourage Edwards and the author to make a similar presentation to City financial institutions and accountants. This event was objected to by the Secretary of State and was allowed to proceed only because it would have been difficult to withdraw invitations already accepted. Edwards and the author were told that this initiative had been unacceptable. Indeed, this episode may well have increased the difficulties in the relationship with senior officials at DEn. On another occasion, Edwards and Parker sought a meeting (with their Chairman's agreement) with John Guinness (the Deputy Secretary with the relevant responsibilities) to discuss the government's line of thinking on future electricity regulation, insofar as it might affect coal. Of particular interest was the extent to which there would be a 'pass-through' of generators' costs (including fuel costs) to the RECs, since limitation of 'pass-through' would have increased pressure from RECs in favour of long-term contracts with predictable prices. At that meeting, DEn officials merely said that they could say nothing, and clearly indicated their contempt for this attempt to raise such matters with them. Indeed, by mid-1989, DEn's position appeared to be that all matters relating to the new ESI contracts were the responsibility of BC and the generators, that there was no question of the government intervening on BC's behalf, and that any attempt by BC to appeal to public opinion would be regarded with great disapproval.

However, as the critical period for resolving the contract issue approached, it became clear that there was no combination of tonnage, price and duration which would be acceptable both to BC and to the private generators, and that no agreement would be forthcoming without government intervention. There had already been signs that the government wished to see a reduction in BC sales to the ESI, in favour of coal imports. In a 'strategy review' in 1988, the Department of Energy had pressed BC to consider a scenario which would have included withdrawal from power station markets distant from the coalfields and from other loss-making markets, effectively encouraging additional coal imports. (BC rejected that option and DEn did not press the point). At that time the government was tacitly supporting a proposal by Associated British Ports (ABP) to build a major new jetty near Immingham on the Humber, capable of unloading 'Capesize' vessels, with a capacity of some 10m. tonnes per annum. Although at the House of Commons Select Committee set up to consider the private Bill (AB Ports No 2 Bill), ABP sought to maintain that the proposal had little or nothing to do with coal traffic, they admitted they had had discussions with the Secretary of State for Energy, and it was widely assumed that the object was to create a large coal-import terminal which would dramatically lower the delivered cost of imported coal into the power stations located in the Central Midlands and Yorkshire coalfields. BC, in the persons of Edwards and Parker, made representations to this Select Committee on the dangers to long-life collieries, including those manned by UDM members (who at that time Mrs Thatcher sought to protect as a reward for opposing Scargill during the Great Strike) and (unsuccessfully) proposed an amendment that import of coal through the new facilities should be prohibited until 1995. Government members (including Mrs Thatcher personally) voted for the Bill which eventually received royal assent in May 1990 (although the terminal was in fact not proceeded with due to the withdrawal of commercial interest by one of the major generators).

The Link with Coal Privatisation

The government could see advantages flowing from ESI privatisation as a means of increasing the commercial pressures on the coal industry. As Margaret Thatcher said in her memoirs: 'Clearly a privately owned electricity industry would be much more demanding in the commercial terms it expected from the NCB (sic) than would a state-owned monopoly' (Thatcher p.685). However, although she stated that 'I always wanted the coal industry to return to the private sector', it was not until November 1990 (just before her fall from office) that she appears to have had any serious discussion about the prospects for full-blown privatisation of British Coal (Thatcher pp.685–6). Indeed, during most of her administration the emphasis had been on ultimately removing the NCB/BC statutory monopoly, rather than on full privatisation. This policy of eroding the NCB/BC monopoly centred on attempts to increase the maximum permitted size for private opencast sites operated under licence. But even on this narrow front, ministers agonised long and hard before making quite marginal changes under the 1990 Coal Industry Act.

Even after the defeat of the Great Strike in 1985, the Thatcher government had been reluctant to declare its intentions on coal privatisation. In 1986, in evidence to the Energy Committee, it said 'The government has no plans to privatise the mining activities of British Coal' (HC 165, para. 173). Mrs Thatcher confirmed this position to the House of Commons on 24 November 1986, and in their response to the Energy Committee's Report in May 1987, the government said 'The priority for British Coal is to restore the industry to financial viability. This is why there are no plans at present to privatise British Coal' (HC 387, para 7.1). An explicit political commitment to privatise the coal industry did not come until October 1988, with the Secretary of State's (Cecil Parkinson) pledge at the

Conservative Party Conference to achieve the 'ultimate privatisation'. Parkinson states his view of the significance of his pledge in his memoirs: 'What was ultimate about the proposed privatisation of coal was that it would mark the end of the political power of the National Union of Mineworkers and would make the coal industry what it should always have been, another important industry, no more and no less important than many others' (Cecil Parkinson: *Right at the Centre*, p.280). Although this pledge earned Parkinson a standing ovation at the Party Conference, and precisely articulated Thatcherite aspirations, it was not seriously acted upon until after ESI privatisation had been secured, and Mrs Thatcher had been succeeded by John Major, who activated work on coal privatisation in January 1991 (John Major: *The Autobiography*, p.248), with the appointment shortly after of government advisors, led by N. M. Rothschild.

This reluctance to press ahead with coal privatisation in the years after the Great Strike reflects once again Mrs Thatcher's caution in coal policy, notwithstanding her enthusiasm for the objective ultimately of returning the industry to the private sector. In spite of the subsequent triumphalist interpretation of the defeat of the Great Strike, the government nevertheless retained some residual anxiety that industrial relations problems might recur particularly if coal privatisation proposals became associated in the public mind with large-scale pit closures. As the Prime Minister said in her memoirs: 'I had learned from hard experience that you must never allow yourself to be manoeuvred into taking drastic action on pit closures when a steady, low-key approach will secure what is needed over a somewhat longer period. In dealing with the coal industry you must have the mentality of a general as much as that of an accountant. And the generalship must often be Fabian rather than Napoleonic' (Thatcher p.686).

Although some influential free-market economists (notably Professor Colin Robinson) had argued that there was a strong case for privatising coal before the ESI, there was never any serious possibility that this would happen. The government recognised that coal industry privatisation would be politically difficult, and certainly would not raise large sums for the Treasury in the way that the ESI did. Moreover, the difficulties of coal privatisation would have been further increased if the future effects of ESI privatisation had remained unknown. Furthermore, privatising the ESI first had certain advantages for the government in relation to coal policy. First, because the complexities of the ESI privatisation would take some time to resolve, there would be more time for the necessary restructuring of British Coal. Second, any further restructuring of BC required before its privatisation could be characterised as the result of commercial decisions by private electricity companies rather than by the government. Electricity privatisation would unleash powerful forces to 'down-size' the coal industry 'by remote control'.

Government Resolves Initial Contract Dispute

We have digressed somewhat in order to provide some background to the circumstances in which the impasse in contract discussions between BC and the new generators was resolved. In order to keep electricity privatisation on schedule, the new Secretary of State (John Wakeham) decided in November 1989 to impose a three-year deal on the parties. For England and Wales, these government-brokered contracts were for three years from April 1990 on a 'take or pay' basis with predetermined prices. Compared with 1989/90, when BC had planned sales of 75m. tonnes to National Power and PowerGen (the two new privatised generators), there was a modest reduction in tonnage to 70m. tonnes in 1990/1 and 1991/2, and to 65m. tonnes in 1992/3. Prices were due to fall by the equivalent of about 4 per cent per annum in 'real' terms. It was clear from this settlement that the government, despite its evident sympathy with the aims of the new generators, would not allow them to use their market power to impose on BC precipitate reductions in coal sales and prices. The new contracts looked like an extension of the 'Joint Understanding' by other means.

The rationale for the way in which the government determined these new contracts appeared to be based on three principles.

Firstly, coal contracts were required to provide an element of price stability in the period immediately following electricity privatisation, particularly in the politically-sensitive domestic market. For this to be done without prejudice to the finances of either the RECs or the generators, such contracts had to be 'back-to-back' into the RECs' monopoly franchise markets under a framework of 'contracts-for-differences' between the generators and the RECs, so that the high costs of UK coal could be passed through to final consumers. At the same time, coal prices needed to fall sufficiently to increase the profitability of the privatised electricity industry while maintaining domestic electricity prices broadly constant in 'real' terms.

Secondly, the coal contracts needed to be sufficiently favourable to BC to avoid the government having to fight a 'second front' on coal until the power stations had been safely transferred to the private sector. In turn, this meant that the new coal contracts had to provide for sufficient volumes of BC coal sales to power stations to avoid large-scale colliery closures which could be attributed directly to ESI privatisation. Further, if the ultimate privatisation of BC was not to be prejudiced, the coal prices in the contracts, although declining in real terms, had to be compatible with BC's progress towards acceptable levels of profitability *without explicit subsidy*. Any 'coal subsidy', which was the difference between BC prices and 'free-market' prices based on parity with imports, would continue to be hidden in the contract coal prices.

Thirdly, there was also the important issue of the duration of the new contracts. Here, the reasoning appeared to be that the contracts had to be of sufficient duration and firmness to preclude any reopening until after the next General Election (due by 1992) and to allow any subsequent contraction of the coal industry to be presented as the result of market forces operated by the privatised electricity, rather than the direct effect of government policy. On the other hand, it already appeared likely that further significant contraction of the coal industry would be required *before* BC could be privatised. Thus, the duration of the new coal contracts could not be so long as to preclude the 'down-sizing' of the coal industry in time to privatise BC within the term of the *following* Parliament. All these considerations suggested a contract duration of about three years.

It is unlikely that the above rationale was ever fully articulated. But whether by accident or great administrative subtlety, the new coal contracts represented a government-imposed short-term reconciliation of policies towards the ESI and the coal industry in 1989/90. Superficially, at least, BC's position appeared promising. The new contracts provided a guaranteed outlet for most of BC's output for three years, at prices well above the prevailing international levels; and the financial reconstruction in the 1990 Coal Industry Act (see Chapter 3 above) provided substantial benefit to the revenue account.

But these measures provided no basis for a secure long-term future for the industry. It was recognised by all parties that both the tonnage and coal prices in the new contracts were significantly higher than free commercial negotiations would have produced, much to the indignation of the generators. Furthermore, the monolithic structure of the contracts meant that the artificial elements of support on both volume and price for 80 per cent of BC's output would expire on a single date: 31 March 1993.

Nevertheless, it can be stated with some confidence that in 1989, when the legislation for electricity privatisation was completing its passage through Parliament, the government had in mind no particular quantum of displacement of BC's power station coal by imports from 1993. Neither did they have any expectation *at that time* that there would in fact be very little additional coal imports, or that instead there would be a huge programme of gas-fired plant which would lead directly to the more than halving of UK deep-mined output by 1994.

Government's Attitude to BC Hardens

However, the government's position towards the coal industry appeared to harden once the ESI privatisation had been secured. The first clear indication was seen in its approach to the Energy (Select) Committee's investigation into the Flue Gas Desulphurisation Programme. In its report (HC 371 published June 1990), the Committee noted that the two privatised generators would probably now install only 8GW of FGD equipment (para 14) and stated that the views of the European Commissioner for the Environment, given in evidence, 'forces us to the conclusion that the Government obtained relatively undemanding limits for the UK on the understanding that the UK would achieve the required reductions [in SO_2 emissions] chiefly through FGD, and that having obtained such limits by that means the UK now proposes to comply with them by cheaper methods instead' (HC 371, para 52). The 'cheaper methods' were imported low sulphur coal or (as we shall see) new gas-fired power stations.

The government's evidence to the Select Committee on this matter was uncompromising, and gave little comfort to BC. The joint memorandum by the Energy and Environment Departments stated that 'The government believes that the electricity generators should be free to follow their own fuel purchasing strategies in the light of all the relevant factors' and that 'the long-term challenge for British Coal is ... to make UK coal the fuel of choice, and FGD the method of sulphur abatement which the generators would wish to choose on economic grounds' (HC371, 1990, Memoranda p.23, para 22 and 23). This signalled that the government was not prepared to protect BC from the effects of the LCPD over and above a diminished FGD programme.

The government's formal response to the Committee's report in October 1990 spelt out its position further (including the rising interest in global warming from about 1988).

> Recent events in the energy sector have underlined the dangers of over-reliance on a single fuel, or source of supply, and the electricity generators fully recognise this. Diversity, in the type of process employed for electricity

generation and in the sourcing of individual fuels provides protection both against short-term disruptions (such as strikes or price shocks) and also against longer-term developments such as the progressive exhaustion of economic UK coal capacity and the need to reduce emissions of CO_2, methane and other greenhouse gases (HC 662, para 33).

These were clear statements that the coal industry was seen as an environmental problem with no offsetting strategic value. Indeed, the government saw a positive benefit in terms of physical and economic security from reducing BC output. The government was also implicitly discounting the claim of '300 years of coal reserves' (which had long been widely quoted as an indication of the UK's physically recoverable coal reserves) as a concept of little meaning in economic terms. By the beginning of the 1990s, environmental issues helped to reinforce the government's view that the UK coal industry was more of a liability than an asset.

It was against the background of such attitudes by government that BC undertook yet another Strategy Review in 1990, designed to address the options available for dealing with the position that would be likely to arise in the mid 1990s. BC calculated that in the crucial power station market in England and Wales an 'optimum level' of BC supply in the mid-1990s would be 56–60m. tonnes per annum at a pithead price of £1.50/GJ (1989/90 money values) – compared with the then contract tonnage of 70m. tonnes per annum and a pithead price of £1.63/GJ. Deep-mined productivity was assumed to increase by 4.5 per cent a year, and colliery operating costs to reduce by 3 per cent per annum in 'real' terms. The corresponding policy options for deep-mined capacity for the mid-1990s were argued to be a 'central' option to retain 50–55m. tonnes and a 'low' option to reduce deep-mine capacity to 35m. tonnes. Here BC saw the conflict between desirability and practicality: although the 'low' option was acknowledged to be the most robust option from a financial standpoint given the environmental and commercial risks, it appeared at the time to present almost insuperable problems of implementation in terms of the scale of further closures and manpower rundown against the government's insistence that BC should continue to follow a policy of 'no compulsory redundancy', and to stay within the constraints of the Modified Colliery Review Procedure.

In preparation for discussions with government on this Strategy Review, the Corporation sought to establish that the best course was progressively to reduce deep-mined output from 76m. tonnes in 1989/90 to around 50m. tonnes by the mid-1990s. At the same time, low-priced markets would be abandoned as soon as existing contracts allowed, there would be continuing pressure to raise productivity and reduce costs, capital expenditure would be reduced and concentrated on existing pits with the best prospects (all thought of new capacity beyond Asfordby having been abandoned), and the possibility of diversifying into coal trade and overseas mining activity would be examined.

Although in discussions with the Secretary of State in June 1990, the government appeared to share BC's view that, in the absence of further government intervention in the market, the 'low' (35m. tonnes deep-mined) option would apply after the end of the new contracts, and that such an outcome was likely to be incompatible with the current instructions from government to avoid 'compulsory' redundancy, the government's considered response to the Strategy Review, received in November 1990, gave little comfort to BC. Some of the main comments made by government were:

- Generally, BC had underestimated the difficulties likely to be faced: BC's assumption of 50m. tonnes per annum to the major generators was likely to be optimistic. The 35m. tonne case was more reasonable. Capacity must be reduced in line with the market, and, more importantly, *the government does not see any particular level below which production capacity should not fall*. (Author's italics)
- On the other hand, BC should not seek to anticipate the end of the existing contracts by under-supplying against those contracts. If necessary, BC should be prepared to buy in supplies from abroad to meet their contract commitments.
- 'The most important event in the next two years will be the negotiations with the generators. The whole future of the industry in the medium-term will hinge on the outcome and it needs to be recognised now that the government will wish to be closely associated with the development of the negotiating strategy as well as the course of the negotiations and will wish to scrutinise the Corporation's proposed settlement carefully before accepting it. However, it must be accepted that by then the government will have no power to affect the generators' position.'
- In due course, BC should seek in the negotiations for contracts beyond March 1993, sales volumes and prices which would secure profitability. It would be particularly important not to seek to buy sales volume at the expense of prices and hence profitability.
- The government believed that, with suitable redundancy terms, and using established closure procedures, it should prove possible to maintain the policy of voluntary redundancy. The Corporation

should not assume that the government would view this matter any differently in future. Moreover, BC should anticipate that the government would wish to see whatever could be done in a cost-effective way to soften the blow to the UDM (as Mrs Thatcher had asked).

BC's senior management reacted with dismay to this government response. A rejoinder was prepared for use in the event of an early rearranged meeting with the Secretary of State, a draft being prepared by the author and circulated to BC's Executive Committee at the end of November 1990. The document noted that BC entirely accepted the financial disciplines of profitable operation within the framework of the current ESI contracts. However,

> so far as life after March 1993 is concerned, the (government) document is largely devoid of hope for the future. Indeed, it is not too strong to say that it shows an indifference to the industry's fate which would be extremely damaging if the document were made public. In our Strategy Review document in June we attempted to address some of these issues but effectively have received no encouragement. Yet these issues have to be resolved if the prospects of a viable British Coal are not to be irreparably damaged by a collapse of morale of management and men, leaving only a 'redundancy culture'.

These intended representations were overtaken by events: Mrs Thatcher ceased to be Prime Minister at the end of November and was succeeded by John Major; and Lord Haslam retired as British Coal Chairman in December 1990, to be succeeded by Neil Clarke.

The Arrival of the 'Dash for Gas'

Once again, the management of the industry was faced with a fundamental problem in which there was an incompatibility between economic and commercial pressures on the one hand, and political constraints on the other. This equation was made the more difficult by the decision of the Major government in early 1991 to quicken the pace of preparation for coal privatisation. As it was by then clear that BC could be privatised successfully only if further major restructuring was completed *before* privatisation, this added a further element of urgency to find a basis for new ESI contracts from April 1993 which would not only be acceptable to the privatised generators, but would provide for profitable operation for a privatised coal industry. Yet, at the same time, as we have already seen, the government were still insisting on continuation of established closure procedures and the avoidance of 'compulsory' redundancy. Above all, the government wished to avoid the issue of large-scale pit closures emerging on the political scene.

This task of establishing politically acceptable phasing to an unknown outcome would, in any event, have represented a formidable challenge for British Coal. By early 1991, however, it was becoming clear that the UK coal industry was facing another serious development in the market: the large-scale move by the privatised electricity industry into gas-fired generation, using Combined-Cycle Gas Turbines (CCGTs): the so-called 'dash for gas'.

During the 1970s and most of the 1980s gas was generally regarded as a 'noble' fuel, to be reserved for 'premium' domestic and industrial uses, and too valuable to be used in power stations, particularly in view of the low thermal efficiency of conventional generating plant. In February 1975, an EEC Directive had restricted the use of natural gas in power stations (75/404/EEC), (although exemptions were allowed where there were strong environmental or economic reasons). The importance of this Directive was not so much as a quasi-legal barrier against gas-fired generation, but as part of the general policy aversion to gas use (shared at the time by UK governments). However, by the end of the 1980s, the more relaxed view being taken on future UK gas reserves, and the emergence of tested CCGT technology (particularly in the USA), made gas-fired generation a practicable alternative to coal.

In their report on the Flue Gas Desulphurisation Programme, the Energy Committee pointed to the environmental and economic advantages of CCGTs, which emitted virtually no SO_2 and only half as much CO_2 per unit of electricity as coal. Also at prevailing fuel prices and using the recently developed CCGT technology, cost estimates at the time indicated that gas provided the cheapest way of reducing SO_2 emissions (HC371, para. 26). The Committee noted that proposals for CCGTs by National Power, PowerGen and independent generators amounted to some 10GW, capable of displacing 25m. tonnes of coal by 2003, although it also noted that not necessarily all of these schemes would necessarily be implemented (HC371, para. 28).

However, the trigger for the 'dash for gas' was not the need to avoid the costs of FGD while complying with the new European SO_2 limits. Rather, the main influence was electricity privatisation, and the particular way in which this was carried through and subsequently regulated. Although it is likely that some gas-fired plant would have been built in any event, nevertheless, once the use of gas in power generation was no longer regarded as undesirable on general energy policy grounds (recognised in the symbolic revocation in 1991 of the EEC Directive discouraging such use), the size and speed of the 'dash for gas' was significantly influenced by the interaction of the new structure of the privatised electricity industry in England and Wales, and the policy of promoting competition in generation. Much of the difficulty of establishing effective competition in generation arose from the treatment of nuclear power before privatisation. We have already noted that, in the White Paper of February 1988, the intention was to include the nuclear stations within the planned privatisation. This meant that one of the new generating companies (National Power) had to be sufficiently large to bear the financial and physical risks of the nuclear component. In turn, this meant that there was room only for one other competing generator (PowerGen), which had to be sufficiently large to avoid being dwarfed by National Power. However, as privatisation approached, the economic and technical risks of nuclear power appeared too great at that time for these stations to be included in the privatisation, and in 1989 they were withdrawn. By then it was judged too late to alter the shape of the National Power/PowerGen duopoly even though the reason for its existence no longer applied.

Although the government's objective was to enhance competition in power generation, it was politically unthinkable for it to contemplate doing so by breaking up National Power and PowerGen so soon after setting them up, or even at that stage requiring them to divest themselves of plant. Thus it appeared inevitable that, for some time, promotion of competition in generation would involve measures that would have the effect of reducing the dominant market shares of National Power and PowerGen, whose generation (like that of their nationalised predecessor, the CEGB) was overwhelmingly coal-based. At the time, the only available route to reducing the market share of the two major generators was through the construction of *new* generating plant owned by neither of these companies; and the most cost-effective way of providing new generating capacity was by building CCGTs, because of their relatively low capital costs and short construction times. In this way, any attempt to increase competition in generation by reducing the market share of the two dominant generators inevitably involved a move into gas at the expense of coal. As the electricity regulator, Professor Littlechild, recognised: (OFFER: *Review of Economic Purchasing: Further Statement*, February 1993, para 80). 'The options presently open to RECs wishing to diversify their supply are extremely limited. In simple terms, they cannot diversify away from their two dominant suppliers without at the same time diversifying their fuel supply [i.e. away from coal], and without bringing new capacity into the market. This is a disadvantage of the present industry structure.' Indeed, if the initial structure of the privatised ESI had provided for (say) five competing (mainly coal-based) generators in England and Wales, it is highly improbable that the 'dash for gas' would have occurred in the way that it did.

But although the policy on competition in generation was the main influence on the 'dash for gas', it was not a sufficient cause. There had to be sufficient incentives for CCGT projects actually to proceed. The RECs saw CCGTs owned by 'independent power producers' (IPPs) as a way to achieve a measure of independence from the two major generators, whose market power they greatly resented: and all the independent CCGT projects in the early years after privatisation had RECs as partners in the consortia. For National Power and PowerGen, the prospect of environmental constraints requiring the expensive fitting of FGD to elderly coal plant made CCGTs attractive, particularly as they might thereby mitigate their loss of market share, and ensure that IPPs did not secure all the available gas supplies.

The development of the CCGT programme was also greatly influenced by the policy of regulation in the gas market, designed to encourage competition by reducing the dominant market share of British Gas, which effectively lost its previous status as monopsony buyer of gas 'at the beach', and at a time when the potential for a substantial increase in gas production emerged. Over the period 1989 to 1993, 36 new gas fields in the North Sea were contracted for, but only nine of these went to British Gas, the other 27 fields being sold to 18 different organisations (M. J. Parker and A. J. Surrey: *UK Gas Policy: Regulated Monopoly or Managed Competition?* SPRU, Nov. 1994, p.38). Under the previous regime, new gas fields were brought into supply only to the extent that British Gas needed them to provide replacement for declining mature fields or to meet incremental demand in British Gas established monopoly markets. As gas competition developed, more gas became available because British Gas was no longer in a position to restrict total supply to the amount needed to satisfy its own forecast demand. The effect was increasing amounts of gas seeking additional secure outlets in the UK (since gas exports were not then practicable). Such markets could be made available by building CCGTs, with the risks covered generally by fifteen-year gas supply contracts 'back-to-back' with contracts for sales of electricity to those RECs who were partners in the projects, and who were able to pass through the costs into their monopoly franchise market (at least to 1998); and the structure of these contractual arrangements enabled the CCGTs to operate at high load factors.

But there was a further element which made the 'dash for gas' possible. As Professor Littlechild stated later: 'Existing and prospective (electricity) prices in the Pool and in the contracts market were high enough to make such investment (in CCGTs) attractive.' (OFFER: *Submission to Review of Energy Sources for Power Stations*, April 1998, para 3.6.) This was in spite of the fact that in the early stages of the CCGT programme, the *costs* of new CCGT generation exceeded those of existing coal stations, as shown by estimates from PowerGen in September 1991, quoted by the Energy Committee in their report on the Consequences of Electricity Privatisation (HC 113, Feb. 1992, para 51). (See Table 4.1).

Table 4.1:   Cost of Electricity from CCGTs and Coal-fired Plant. PowerGen Estimates September 1991. p/kWh

	Without FGD	With FGD
New CCGT: gas at 23p/therm	2.89	-
New CCGT: gas at 20p/therm	2.64	-
Existing inland coal, using British coal	2.20	2.73
Existing inland coal, using imports	1.66	2.19

Source: HC113 Table 4 (para. 51)

Although the case was less clear cut in cases where FGD was required, it was clear that given the prevailing coal and gas prices in 1991 *existing* coal stations generated more cheaply than *new* CCGTs, after taking account of the capital charges on new capital expenditure on the gas-fired plant. The reason that this did not prevent CCGTs being built was that the decisions were based on *prices* not costs. As Professor Littlechild was to say:

> From the point of view of a supplier purchasing electricity contracts, the relevant consideration is the price at which electricity is offered over the contract period, rather than the costs of the generator in producing it. It might have been possible for the coal generators to bid into the Pool or offer contracts for differences at prices which sought to maintain output of coal stations and made new entry by the first gas generators unprofitable or at least significantly less attractive. *They did not do so.* (Author's italics) They may have considered that it was more profitable for them to maintain prices than to compete more aggressively for market share ... It is possible that, had the generation market been more competitive at Vesting, and had British-produced coal been more competitively priced and available from several producers rather than just one, the incentives on the incumbent coal-fired generators would have been different. But this was not the case (Ibid 3.9).

On the other hand, in circumstances when the Regulator was seeking to reduce the market share of the major generators in any case, they had little incentive to try to maintain their market share by reducing prices.

The interaction of all the above factors led in fact to a very rapid rate of ordering of CCGTs. By the end of 1990, some 4GW of gas-fired plant had been ordered, 6GW by the end of 1991, and 11GW by the end of 1992, by which time it was virtually inevitable that, because the plants were planned to operate at high load factors and could be built quickly, over 30m. tonnes of BC's market in power generation would be lost by the mid-1990s.

BC's Interlocking Problems

Prior to ESI privatisation, it was generally assumed, both by the government and by BC, that the main threat to BC's sales volume would arise from coal imports: whereas by 1991 it was clear that the main threat to the size of BC's market arose from the 'dash for gas'. However, this did not mean that trends in international coal were no longer of major concern to BC.

The main issue here was price, since to become commercially viable, BC would have to be competitive with internationally traded coal even if the UK coal market was reduced by the 'dash for gas'. We have already noted that, in real terms, the $ price of international coal imported into the EU halved over the 1980s, and in the early 1990s this real price reduction was sustained. In sterling terms, there was further downward pressure arising from the stronger £ in relation to the dollar. The 4-year moving averages showed a continuing downward trend in prices. (See Table 4.2).

Thus, from 1991 BC was faced with conducting interlocking discussions with the government on the industry's forward strategy,

Table 4.2: 'Real' Prices of Imported Coal for Power Stations in the EU

	4 year Moving Average 1998 $/GJ	4 year Moving Average 1998 £/GJ
1989	2.13	1.52
1990	2.12	1.35
1991	2.13	1.32
1992	2.11	1.33
1993	1.99	1.24

with N.M. Rothschild on preparation for coal privatisation, and with the generators on replacement coal contracts after 1992/3, against the background of escalating future losses of market share as the 'dash for gas' accelerated, and a continuing downward trend in the price of internationally traded steam coal.

Although initially DTI officials had expressed scepticism on the scale of the 'dash for gas', by early 1991 BC had come to recognise that the end of the coal contracts with the major generators in March 1993 would be followed by significant reductions in both sales and prices, and that there would be a substantial associated fall in deep-mined output. In July 1991, the Corporation submitted yet another Strategy Review to government which put forward two cases for the five years 1993/4 to 1997/8 (see Table 4.3).

Table 4.3: Alternative BC Scenarios for Coal Contracts as Seen in 1991

	Sales to Major Generators (England and Wales)	Deep-mined Output
Case I	40 m.t. 1993/4, falling to 35 m.t. p.a. to 1997/8 Average price £1.50/GJ (1991 money)	35 m.t. in 1993/4, falling to 30 m.t. p.a. to 1997/8
Case II	50 m.t. in 1993/4, falling to 40 m.t. in 1997/8. Price falling from £1.70 to £1.50 (1991 money)	45 m.t. in 1993/4 falling to 35 m.t. in 1997/8

At the time of the submission, sales to the ESI in England and Wales were 70m. tonnes, at a price of about £1.80/GJ and deep-mined output was over 70 m. tonnes. As early as the first half of 1991, therefore, BC's analysis indicated, on the basis of unchanged policies, a virtual halving of deep-mined output and sales to the generators in the period following the expiry of the then current contracts, together with large reductions in price.

This did not mean that BC necessarily advocated this kind of outcome. Indeed, there was some difficulty in deciding what position BC should take in these strategic discussions. Although the discussions with N.M. Rothschild on the pre-privatisation analysis had begun in May 1991, there was then no clear idea what form BC's privatisation would take. This placed BC in something of a dilemma. Was it still planning its own future, or the future of other entities whose existence had not yet been determined? Was its task at that point to plan for the largest 'economic' industry post privatisation, or should greater emphasis be placed on financial robustness rather than size? This question impinged on the attitude to be adopted to the trade-off between future volume and price, particularly in any successor contracts with the generators. In short, should BC be planning to maximise profitability on risk-averse assumptions, or was it trying to secure the largest possible industry consistent with likely financial, commercial and environmental constraints?

There were also ambiguities in the government's attitude to the industry in 1991. Before the submission of the Strategy Review in July 1991, BC had been told by the Department of Energy that they did not want any firm conclusions about the terms of the successor contracts before the next General Election. As a further year might elapse before the Election, such a requirement was likely to increase problems later. As BC said in their Strategy Review:

> The Corporation would not wish to move to the rapid rate of contraction implied in Case I until the contract with the generators is established, as doing so would involve a large closure programme which would be difficult to achieve. If negotiations of the contract is (sic) delayed until after the next election, British Coal might be faced with the generators having entered into arrangements for supply of alternative fuels which could exacerbate our problems. We will be seeking to put together a framework of contracts with the generators and the [electricity] distribution companies which will achieve the maximum size for a sustainable U.K. coal industry consistent with a guarantee to reduce electricity price for [domestic] franchise customers, together with a more acceptable rate of contraction if that proves to be necessary. It is important that this issue is brought to a speedy resolution in order that the generators do not take actions now which would prevent the above objective being achieved.

BC also warned (prophetically, as it happened) that if both successor contracts and consequent BC colliery closures were to be delayed until after the next Election, and 'if we are forced to a market size similar to Case I the newly elected administration would face the prospects almost upon taking office of seeing the contractual arrangements with the generators and resulting closure programme between say June 1992 and March 1993 which is bigger in proportion to anything yet achieved'.

In his reply in September 1991 to BC's 1991 Strategy Review, the Secretary of State (John Wakeham), recognised that a major closure programme would be needed, although its size would not be clear until contract negotiations were completed. But 'the negotiations will be conducted on commercial terms and the contracts will be settled at market prices. The government's role will therefore be very limited.' There was no indication that the government would seek in any way to slow the 'dash for gas', and consents for new gas stations continued at a rapid pace.

Origins of the October 1992 'Coal Crisis'

Although BC again warned the government in November 1991 of the colliery closure problems likely to be faced if the end of the ESI contracts led to sharp contraction, and in particular the impossibility of preserving any semblance of a policy of 'no compulsory redundancy', in the months before the April 1992 General Election, the government's position on coal was to do nothing to stand in the way of the forces which would put strong downward pressure on BC's markets and prices after March 1993. At the same time they wished as far as possible to avoid any public controversy over future closures and manpower rundown until (as they hoped) the government had been returned in the forthcoming General Election. The government's aversion to such controversy was strengthened by the leak in September 1991 of the first interim report by N.M.Rothschild indicating the likely future scale of closures, even though the Department of Energy's press release described this as no more than speculation.

The tactic of putting all solutions into suspense until after the election made a 'cliff-edge' crisis later in 1992 almost inevitable, assuming the Major government was returned. It also made for considerable difficulties for BC in deciding what its public position should be. At a time when BC's senior management was becoming increasingly alarmed at the scale of the problems ahead, government was above all concerned that BC should not 'rock the boat' by giving public expression to its fears in an effort to attract political sympathy. In these circumstances, BC attempted a balancing act whereby it drew attention to the dubious economics of the 'dash for gas' (as the BC Chairman, Neil Clarke, did on a number of public occasions in 1991/2), while at the same time refraining from spelling out the full consequences in terms of closures and lost miners' jobs. The unsatisfactory nature of this compromise, which was a direct result of the nature of BC's dependent relationship with the government, was shown in BC's approach to the Energy (Select) Committee's inquiry on the Consequences of Electricity Privatisation (Report HC 113, Feb. 1992). When BC witnesses gave evidence in December 1991, their reluctance to set out the 'worse case scenarios' was badly received. By contrast, the separate evidence of Malcolm Edwards (BC's outgoing Commercial Director) spelt out how few collieries would survive in the absence of policy changes. His evidence was welcomed by the Committee: 'In our view, Mr Edwards has performed a major public service in demonstrating the consequences of present policies within the electricity supply industry and the fact that the country will need to decide very quickly whether it wishes to retain a significant indigenous coal industry' (HC113 para 150). But notwithstanding the warnings of the Select Committee, published on 26 February 1992, the political boat was not rocked. Policy still remained on hold until after the General Election in April 1992 (which saw John Major returned, but with a much reduced majority).

In contrast to BC's increasingly pessimistic (but private) analysis of the industry's future, the actual performance of the industry in 1990/1 and 1991/2 in other circumstances would have led to optimism and congratulation.

Rationalisation continued against a background of modest reductions in output and sales.

Over the two years 1990/1 and 1991/2, BC reduced both the number of operating collieries and colliery manpower by about a third, and labour productivity increased by 26 per cent. (See Table 4.4)

There was also considerable improvement in the financial results,

Table 4.4:   BC Output, Sales and Productivity. 1989/90 to 1991/2

	1989/90 *	1990/1	1991/2
Output (m.t.)			
Deepmines	75.6	72.3	71.0
Opencast	17.5	17.0	16.7
Licensed mines	2.1	2.3	3.4
	95.2	91.6	91.1
Sales (m.t.)			
Power stations	75.8	74.3	73.0
Other markets	18.5	16.0	13.4
	94.3	90.3	86.4
Number of collieries at year end	73	65	50
Colliery manpower at year end ('000)	65.4	57.3	43.8
Output per man year	1080	1181	1357

*   1989/90 was a 53 week year.

Source: *BC Annual Report 1994/5*

with net profits, after interest and exceptional items, of £78m and £170m in 1990/1 and 1991/2 respectively. (See Table 4.5). The emergence of net profits owed much to the Financial Reconstruction in the 1990 Coal Industry Act, and the associated reduction in capital charges, which were reduced by over £600m a year (See Table 4.6).

BC thus gained very substantially from the more realistic valuation of assets. Nevertheless, the results in 1990/1 and 1991/2 represented a real achievement, with costs being reduced by greater efficiency.

By the time the Major government was returned to power in April 1992, average colliery operating costs were some 40 per cent lower in real terms than in 1982/3 (the last year before the Great Strike), of which only about 6 per cent was attributable to the reduction in depreciation due to the financial reconstruction; and average opencast costs had also been reduced by some 25 per cent in real terms over the same period. Since the end of the Great Strike, colliery manpower

Table 4.5: BC Financial Results 1990/1 and 1991/2. £million

	1990/1	1991/2
Operating Results		
Deep mines	43	153
Opencast	150	171
Others	45	37
Total operating profit	238	361
Interest	(143)	(93)
Exceptional restructuring items*	(17)	(98)
Net profit	78	170

* Net after receipt of Government grants

Source: *BC Annual Reports*

Table 4.6: BC Capital Charges 1989/90 to 1991/2. £million

	1989/90	1990/1	1991/2
Colliery depreciation	345	173	156
BC interest charges	574	143	93
Total	919	316	249

Source: *BC Annual Reports*

had been reduced without compulsory redundancy from 171,000 to 44,000, and output per man-year increased from 504 tonnes in 1982/3 to 1357 tonnes in 1991/2.

Yet, although the progress of radical restructuring continued in 1991, the achievements of BC in this respect, which had previously attracted fulsome praise from government, were no longer so regarded once the momentum towards coal privatisation increased. The Department of Energy commissioned a study by Alan Oakes, previously General Manager of BP Coal in Australia. The report ('Deep Mine Productivity Comparisons': Alan Oakes and Associates, March 1991) was in some important respects critical of BC performance. The report recognised that UK mines were often working highly faulted areas of coal which would be avoided in the USA and Australia, and that the attitudes of the Mining Inspectorate and the trades unions had inhibited greater use of new techniques on mistaken concerns about safety. On the other hand, the 'Oakes Report' criticised BC for over-centralisation of organisation and slowness to adopt alternative technology, including 'bord and pillar' working and rock-bolting for roof supports. The report concluded that, although BC productivity had almost doubled over the previous five years, this was from a low base, and that the productivity gap compared with USA and Australia was widening as those countries continued to press ahead with new technology.

The 'Oakes Report' was followed by a further study by the American mining consultants, John T Boyd Company, as part of the Rothschild privatisation analysis. Boyds assessed the scope for cost reduction at 28 collieries which were generally thought to have a long life. Their report in April 1992 distinguished between those cost reductions which might be achieved either assuming working practices permitted under existing legislation on the length of the working shift, and given the position of the Mines Inspectorate on supervisory structure at collieries; or assuming that these obstacles were removed. In practice, the Boyd estimates were similar to those of BC, which envisaged a continuation of cost and efficiency improvements.

Certainly, substantial cost reductions would be required if BC was to achieve competitiveness with international coal. In 1990/1 and 1991/2 both BC prices and average colliery operating costs remained substantially higher than the equivalent price of imported steam coal. Although a precise comparison is complicated by differences in quality (with BC coal generally having higher sulphur and chlorine content) and by transport costs within the UK, a broad indication can be obtained in terms of the cost/price per GJ, and by adding some 15p/ GJ to the import price as a broad estimate of the *average* transport cost advantage of BC coal in UK markets at the time. On this basis, over the two years, BC average colliery operating costs were over 40 per cent higher than the notional import price, and BC average prices were 50 per cent higher than the import price (See Table 4.7). In terms of the fundamentals, this puts a different perspective on BC's profitability. In 1991/2 BC's deep-mined operating profit margin was only 5 per cent in spite of the very large price premium. The industry was still a long way from fully commercial operations at market prices.

Table 4.7: UK Deep-mined Costs and Prices, Against Imported Prices 1990/1 and 1991/2. £/GJ

	Average Colliery Operating Costs	Average Colliery Revenue	Import Prices (plus 15p)
1990/1	1.62	1.64	1.09
1991/2	1.57	1.69	1.15

In addition to the covert price subsidy in the ESI contracts, BC also continued to be heavily dependent on government for external finance, particularly for social cost grants to facilitate restructuring and manpower rundown within the objective of 'no compulsory redundancy'. In 1991/2 these grants totalled £457m.

Thus, in spite of all the actions taken to increase efficiency and close uneconomic pits, when discussions on the industry's future resumed following the General Election in April 1992, BC's position remained fundamentally weak. The Corporation was dependent, both upon the privatised generators to secure acceptable contracts from March 1993 (at a time when the balance of market power was overwhelmingly against BC), and upon the willingness of government to continue to finance the restructuring and redundancy programme. BC therefore made it clear to government that no alignment to the much lower sales volumes arising from the likely effects of the 'dash for gas' could begin *until* the shape of the new ESI contracts was known; and, secondly, that, given the scale of the likely adjustment, new and more generous redundancy arrangements, paid for by government, would have to be agreed *before* any announcement on a large closure programme could be made. Any delay in settling either of these issues would change a potential 'cliff-edge' problem into a crisis. But that is what happened.

The Problem of Negotiating New Contracts

From the outset, the extreme difficulties of the contract negotiations were apparent. But during 1992, these difficulties increased. The market prospects continued to decline, particularly as the plans for new CCGTs continued unabated – between January 1992 and October 1992, a further 4GW of capacity received government consent, so that, by the latter date, potential CCGT capacity was nearly 13 GW, which could displace nearly 40m. tonnes of coal use in power stations if all the stations with consent were to proceed. But there was another factor, which was to prove as important as the 'dash for gas' in eroding BC's sales in the two years after March 1993. Coal stocks at power stations were high and rising; reaching 32m. tonnes by October 1992. Such stock levels were greatly in excess of commercial needs, particularly given the prospect of falling coal consumption, and further increased the market power of the generators in relation to BC. A 'draw-down' of coal stocks was now a major alternative low-cost fuel supply in substitution for BC current output: an option which National Power and PowerGen showed every intention of exercising once the existing BC coal contracts expired in March 1993. Finally, it was increasingly clear to the major generators and BC that any compromise, especially one that involved prices significantly higher than alignment with international coal prices, would have to be 'back-to-back' with contracts between the two major generators and the twelve Regional Electricity Companies, with the higher cost of BC coal being passed through to the RECs' monopoly franchise (domestic) customers until 1998. This was a formidable task, and it was difficult to see how it could be accomplished without government co-ordination. For its part, the government had an interest in ensuring that the outcome would be compatible with the subsequent successful privatisation of BC, and without great political controversy over pit closures and redundancies.

From December 1991 BC made a number of offers to the two generators which embodied a very substantial reduction in both tonnage and price. The offer in May 1992 is shown in Table 4.8. But it soon became clear that even this proposal would not be acceptable either on price or volume. The generators also reported that, in their own discussions with RECs on contracts for coal-based generation, the volumes on offer fell well short of BC's aspirations. The aim of Tim Eggar (the new Energy Minister within the reconstituted DTI following the abolition of the Dept. of Energy after the Election) was to settle the contracts in June, but this soon became unrealistic.

Table 4.8: BC Contract Offer to Generators May 1992

	1992/3 (Existing Contract)	1993/4	1994/5	1995/6	1996/7	1997/8
M. tonnes	65	50	40	40	40	40
Price/GJ (1991 money)	£1.77	£1.55	£1.51	£1.47	£1.43	£1.40

The government made some attempt to resolve the impasse without becoming directly involved in the negotiations. Keith Palmer, the head of the Rothschild team on Coal Privatisation acted as go-between in an intensive series of meetings between generators and RECs, and with some involvement of BC – but to no avail. About this time, a series of future BC contract prices, not offered at that time by BC, began to be circulated as 'DTI guidance figures', which envisaged prices of £1.50/GJ in 1993/4, falling to £1.30/GJ in 1997/8. Although DTI denied that these figures had official standing, they provided a framework for discussions between the parties, and gained credence in a way that increased the pressure on BC to make further price concessions.

At the end of July 1992 Neil Clarke (BC Chairman) had discussions with Tim Eggar on the failure to reach agreement on the contracts. At this meeting, the Minister indicated that the government would be prepared to re-examine some of the details of the structure of the privatised electricity industry, if this was necessary to maintain a sizeable UK coal industry. However, a subsequent letter from the DTI again asked BC to submit proposals by early September on the scope for further price reductions.

These matters came to a head on 3 September 1992, at discussions between BC and Michael Heseltine and Tim Eggar, at which BC was asked to make a further price concession along the lines of the informal 'DTI guidance figures'. It was also implicit that sales volumes would be in line with the latest advice from Keith Palmer of Rothschilds that the maximum contract tonnages would be 40m. tonnes in 1992/3 and 30m. tonnes in each of the following four years. Accordingly, Neil Clarke sent to Tim Eggar on 7 September a revised price offer and after acceptance by Eggar on 8 September sent a final offer to the generators in line with government advice on the best that was negotiable. (See Table 4.9)

In his letter to Eggar the BC Chairman had spelt out the consequences of such an offer, which included the closure of 32 deep mines in 1993/4.

Table 4.9:   Final BC Contract Offer to Generators September 1992

	1992/3 (Existing Contract)	1993/4	1994/5	1995/6	1996/7	1997/8
M. tonnes	65	40	30	30	30	30
Price/GJ (Oct. 1992 money)	£1.83	£1.51	£1.46	£1.41	£1.37	£1.33

BC and Government Arrive at the Cliff-edge

By the beginning of October, BC's final offer had been neither accepted nor rejected by the generators, as considerable complications still remained in settling the associated contracts between the generators and the RECs. However, following the discussions with government, BC now understood that negotiable contract volumes would *at most* be those contained in their final offer.

While the coal contract negotiations had followed their tortuous course to October 1992, parallel discussions on new redundancy arrangements also followed an agonisingly slow path. We have already noted that the Redundant Mineworkers Payment Scheme (RMPS), which had been directly funded by government, ended in March 1987, and was followed by four successive BC schemes, which were funded by government via their contribution to Restructuring Grants. The fourth scheme, which covered over 26,000 redundancies, at an average lump sum payment of £24,000, effectively came to an end in March 1992, when the special 'British Coal' supplement of £10,000 included in the scheme was withdrawn, and could not be reinstated without the agreement of government. Yet the widespread expectancy in the coalfields of a new and more generous arrangement effectively prevented any action on colliery closures until a new scheme was in place. By June, it was already clear that BC faced an unprecedentedly sharp reduction in capacity, and BC briefed the DTI that it could involve at least 25,000 men losing their jobs in a single year (*BC Annual Report 1992/3*, p.8). BC therefore argued not only that the Modified Colliery Review Procedure should be replaced by a new speedy process, but that the previous redundancy scheme should be continued, and with enhanced payments. Discussions on these proposals made slow progress, with protracted disagreement between DTI and the Treasury (Heseltine, *Life in the Jungle*, p.437). BC was not told of the government's decision until 5 October, that the same terms of the March 1992 scheme were to be reinstated, but with no further enhancement. Moreover, the Treasury made it clear that 90 per cent of the government expenditure on the new redundancy arrangements would have to be incurred by March 1993.

By early October, therefore, it was clear not only that BC was faced with a very large reduction in capacity by March 1993, but that the closures would have to be announced simultaneously since it was barely possible to complete the programme in the time allowed. Further, it was not possible to enter into statutory consultations (for a minimum of 90 days) if the new terms were to be paid to redundant mineworkers (*BC Annual Report 1992/3*, p.9). Moreover, the leak to the NUM and the press of a letter from Tim Sainsbury (Industry Minister) to Michael Portillo (Chief Secretary to the Treasury), clearly indicating the scale of the closure programme to be expected, had persuaded the government that, once the redundancy terms had been agreed, BC should make a comprehensive announcement on the closures involved, without any pretence that these were related to consideration of the collieries' performance under the Modified Colliery Review Procedure.

British Coal and the government had arrived at the cliff-edge.

On 12 October 1992, at a meeting to discuss the terms of BC's announcement, Neil Clarke emphasised to Michael Heseltine that an extremely hostile response was to be expected. On 13 October, the announcement was made (BC Press Release 2727) that:

> British Coal is to cease production at 31 of its 50 deep-mine collieries because of the reduced demand for its coal. Twenty-seven of the collieries will close. The remaining four will be kept on care and maintenance to maintain access to reserves.
>
> The decision is in the light of harsh conditions in the electricity market and the urgent need to bring supply and demand back into balance. While negotiations with generators serving England and Wales for five-year coal contracts from next April (1993) continue, it is now clear that the maximum volume which can be obtained in the first year (1993/4) is likely to be 40 million tonnes, compared with 65 million tonnes in the current year.
>
> Colliery closures will take place during the coming five months. Most will cease production in the next few weeks. Up to 30,000 mineworkers and staff will lose their jobs… The future of the remaining collieries will depend on the detailed supply arrangements to be agreed with the generators, National Power and PowerGen, and on colliery costs.

The BC announcement also gave details of the redundancy terms agreed with government, with benefits depending on length of service and weekly earnings, up to a maximum of £37,000, but pointed out that the 'payment of maximum redundancy entitlements will depend on the excess capacity being closed quickly'. BC also made it clear that there was no realistic prospect of the Corporation maintaining its voluntary redundancy policy; compulsory redundancies would now occur. Further, the Modified Colliery Review Procedure, which had hitherto allowed appeals against closure, would be set aside.

Inspite of the (genuine) sentiments of Neil Clarke, the BC Chairman, that 'the closure of so many of our mines and so many jobs will be grievous,' and that 'everything possible must be done to ease the difficulties which will be faced by those leaving British Coal, by their families and communities', the starkness of BC's decision announcement caused a great outcry, not only in the coal industry, but throughout the country. Public opinion was overwhelmingly directed against the government, rather than British Coal.

British Coal's announcement had come at a critical time for the government, with its authority already weakened in September by 'Black Wednesday', when Britain had been forced out of the European Exchange Rate Mechanism in humiliating circumstances. The coal industry, which, since the defeat of the Great Strike, had seemed to have lost most of its political significance, suddenly became the focal point for general dissatisfaction with the government. The social impact on the mining communities, which had long had a special place in the national consciousness, was strongly emphasised by the media and Church leaders. As John Major was later to write; 'The prospect of whole mining communities being destroyed touched a raw nerve among the British people – including grassroot Conservatives, who understandably felt we had a moral obligation especially to the Nottinghamshire miners in the Union of Democratic Miners (sic) who had continued working throughout the long coal strike of 1984–85. The Conservative benches were as angry as Labour's' (Major p.669). With sufficient Conservative backbench rebel MP's (including disaffected 'Eurosceptics') to threaten the government's modest majority in the House of Commons, the 'coal crisis' became a major political issue.

The government's initial public reaction was to claim that the BC closure announcement was the result of market forces, and that there was no case for government intervention. However, this position rapidly became untenable. On 19 October, following an emergency Cabinet meeting, Neil Clarke was summoned to meet Michael Heseltine and was told that the government could not proceed with the closures as announced because the political opposition was uncontrollable. Instead, BC would need to have a cooling-off period when only voluntary redundancy would generally apply, and BC would be able initially to close only a limited number of pits. That day, Heseltine announced that the government had introduced a moratorium on the closure of 21 of the 31 pits which had been announced for closure the previous week; and on 21 October, he announced to the House of Commons that the government intended to undertake a wide-ranging review of their prospects.

The government had badly mishandled an inherently difficult situation, as John Major later admitted (Major p.670); and Michael Heseltine later wrote, 'Mistakes were made in misjudging the public mood both in its preparedness for the [closure] announcement and following on from the ERM crisis. There is no doubt either that, if Norman Lamont and Michael Portillo at the Treasury had not simply dug in when the problem was first put to them, the closures would have taken place within an earlier and more orderly programme' (Heseltine p.444). But there were other considerations. There had been a failure at an early stage to anticipate the scale of contraction of deep-mined output and manpower from March 1993 that would result from the cumulative effects of government consents for new gas-fired power stations, and the major generators' wish drastically to reduce their coal stocks. And the potential 'cliff-edge' character of the problem had been exacerberated by delays to contract negotiations arising from the 1992 election campaign, as well as by the late timing of decisions on new redundancy arrangements. As a result the UK coal industry had arrived at another critical point in its history under the full glare of political controversy.

CHAPTER 5
THE COAL REVIEW AND THE SETTLEMENT OF MARCH 1993

Process of Government's Coal Review

Following the announcement by the President of the Board of Trade (Michael Heseltine) to the House of Commons on 21 October, that there would be a moratorium on the closure of 21 of the collieries announced for closure by BC on 13 October, while a review of the closure programme was carried out, BC was subject to five months of intense and highly politically-charged scrutiny of its performance and prospects under three interlocking examinations, namely:-

- First, the government's own Coal Review, which was concluded with the publication in March 1993 of the White Paper 'The Prospects for Coal: Conclusions of the Government's Coal Review' (Cm. 2235).
- Second, the Trade and Industry (Select) Committee (TISC) at the same time began an enquiry to 'consider the consequences of British Coal's pit closure programme for the electricity consumer, the Exchequer and the economy, and to examine alternatives in terms of energy policy'. The TISC report 'British Energy Policy and the Market for Coal' (HC 237) was published on 26 January 1993.
- Third, the Employment (Select) Committee conducted an enquiry resulting in their report on 'Employment Consequences of British Coal's Proposed Pit Closures', published on 13 January 1993 (HC 263).

The terms of reference for the government's Coal Review, published on 26 October 1992, made it clear that the review of the future of the 21 pits subject to the closure moratorium would be wide-ranging, and would be conducted 'in the context of the Government's energy policy, including the consequences of that policy for British Coal and the employment prospects for the industry. It will decide whether the case for closure at each of the pits in question has been fully made and whether it is sensible to mothball some of these pits. It will consider whether the market prospects for each have been correctly assessed.'

The review would also 'explore the opportunities for the private sector in the production of coal', and consultants would be asked to report, inter alia, on 'the competitiveness of British Coal as an organisation'. (Appendix A to Cm.2235).

The government's motivations in the Coal Review were clear: to defuse the political and public outcry against the closure announcement of 13 October; to show as far as possible that the scale of the proposed closures was primarily the result of market forces outside the government's control; that to the extent to which this was not the case, blame should rest with British Coal rather than the government; and finally that the outcome should not prejudice the timetable for the privatisation of British Coal. The objectives of the government's opponents on the coal issue were the opposite: namely, to show that the 13 October closure announcement had been the direct result of government policy, that the potential market for UK deep-mined coal could in fact be significantly increased if appropriate measures were taken, and that (in the view of most of the government's opponents) BC privatisation should be abandoned.

Before the work on the Coal Review had properly begun, there were legal complications. Three of the Mining Unions (NUM, NACODS and UDM) joined action with individual mineworkers in litigation against British Coal and the government (in the person of the President of the Board of Trade), seeking judicial review of the decision of 19 October 1992 to close the ten collieries excluded from the closure moratorium, without resort to the Modified Colliery Review Procedure (MCRP) or an alternative consultation procedure to the same effect; the decision of British Coal on 13 October to cease production at these ten collieries; the decision of the President not to instruct BC to consult with the Unions; the decision of the President not to include the ten collieries in the general review which covered the other 21 collieries which had been included in the original closure announcement of 13 October; and the decision of the President not to instruct BC to refrain from ceasing production pending consultation.

Judgement was given by Lord Justice Glidewell on 21 December 1992, which concluded that the decisions in the announcement of 31 closures on 13 October and of ten (out of the 31) on 19 October were unlawful, in that 'all ignored British Coal's obligations under Section 46 (1) [of the Coal Industry Nationalisation Act 1946] and completely failed to satisfy legitimate expectation of the mineworkers' unions and their members that the MCRP would continue to be followed unless and until notice to the contrary had been given' (Glidewell Judgement p.58). The judgement also declared that 'British Coal shall not reach a final decision on the closure of any of the ten collieries, nor shall the President make available funds which would enable British Coal to reach such a decision, until the procedures substantially to the same effect as the MCRP including some form of independent scrutiny had been followed in relation to each of the collieries.' Following the suggestion of Lord Justice Glidewell, the point on 'independent scrutiny' was met on 23 December 1992, when Michael Heseltine appointed the American Mining Consultants, John T Boyd, to examine the ten pits in addition to the 21 pits included in the closure moratorium on 19 October.

The Glidewell judgement was hugely embarrassing both to British Coal and the government. However, the practical effect was limited. Effectively the closure of the ten collieries had been suspended during the Court proceedings; and the outcome was that all the 31 collieries included in the original closure decision of 13 October were now being formally reviewed. Moreover, as the Coal Review proceeded, it became clear that the future of *all* BC's 50 collieries was being reviewed, since only in that way could proper account be taken of the overall market constraints, and a proper assessment made as to whether BC had adopted appropriate criteria for closure.

The central issue of the government's Coal Review (and the parallel investigations of TISC) was the future size of the market for UK deep-mined coal. In its formal submission to the Coal Review (1.12.92), BC stated that its principal objective was to achieve 'the largest economic United Kingdom coal industry in the longer term' – a formulation based on a statement by the Secretary of State for Energy (John Wakeham) on 17 October 1991, in a written answer in the House of Commons (para. 1.3). To this end, BC stated that 'many of the policy initiatives which we suggest in this paper are designed to correct what may be judged as distortions in the market place. We believe that action should be taken to increase the market for coal in order to allow the UK coal industry sufficient time to complete the transition to market competitiveness' (para 1.4). This theme – that the coal industry should be given more time to make itself competitive and to allow adverse market distortions to be removed – was to remain a part of 'the case for coal' in the years ahead. During the 1992/3 Coal Review, these matters were debated, not merely in terms of the optimum economic outcome, but against the background of politically polarised partisanship.

More specifically, BC's submission to the Coal Review stated that

British Coal believe that the potential exists to correct current distortions in the market place and take other actions to increase the UK market for coal in 1997/8 (i.e. the last year of the proposed 5-year contract period) by up to some 20m. tonnes per year and to enable the generators to enter into term contracts to buy it. Realising this potential depends on initiatives which lie almost exclusively within the power of government (para 2.6).

Changes in electricity regulation would enable the generators to contract with BC for 15m. tonnes of coal per annum in addition to the volumes in the contracts then under discussion (para 2.14); and this additional coal could be supplied profitably provided the government took legislative action to 'release the industry from the current severe restraints in making changes to working practices' (para 2.9). BC's contention was that, if these measures were implemented, the previously-planned 'cliff-edge' of closures could be avoided, with 32 pits still operating in March *1994*, (whereas previously the expectation had been only 19 pits operating in March *1993*), and with average colliery manpower over the period to 1997/8 some 7,000 a year higher than it would otherwise have been.

BC concluded that 'an expansion of the market for coal by the Government action we are proposing would therefore have substantial benefits for the economy, the coal industry and for jobs'. There is a strong suggestion here that the crisis of October 1992 could have been greatly mitigated (particularly the 'cliff-edge' character of the closure programme and associated job losses) if the government had adopted different measures. In other words, it was implicit in BC's submission that much of the problem arose from a lack of political will to solve it, rather than the overwhelming inevitabilities of the market. Moreover, BC's submission implied that there was still time for remedies to be applied. Clearly, there were difficulties for BC engaging in a public and political process. As a nationalised industry, it was not open to BC directly to criticise government policies, but it came close to doing so. While it could be argued that BC was doing no more than making assessments of the consequences of 'unchanged policies', and indicating what measures were available if other outcomes were required, this is unconvincing: in its submission BC was *advocating* a larger market for UK deep-mined coal.

BC's submission concentrated on the scope for additional sales to the electricity generators in England and Wales. Its suggested policy initiatives available to government, which could increase potential sales by 20m. tonnes by 1997/8 are shown in Table 5.1. In addition BC proposed that

- The government should consider retaining higher minimum 'strategic' stocks at power stations and to allow the proposed rapid

Table 5.1: BC Suggestions for Increasing Coal Sales to Generators (England and Wales)

	Increase in Market for Coal in 1997/8 (m. tonnes)
not extending licences of Magnox nuclear stations	5
deferring commissioning of Sizewell B nuclear station	3
reducing gas burn at 'committed' CCGTs	3
embargo on construction of further CCGTs	3
eliminating net imports of electricity through the French link	6
Total	20

reduction of coal stocks by the generators to be spread over a longer period, to mitigate the reduction in sales of current output expected in 1993/4 and 1994/5.

• the amount of Orimulsion used in power stations should be limited to that covered by existing consents.

• the RECs' existing franchise limit should be maintained at 1MW from April 1994 and the franchise period extended beyond March 1998, to facilitate the negotiation of larger tonnages of coal in 'back-to-back' contracts between British Coal and the generators, and the generators and RECs; together with a review of whether the RECs might reasonably enter into longer-term coal-based contracts beyond 1997/8, whether or not the franchise period was extended.

These proposals, or variants of them, formed the basis of most of the discussion in the Coal Review on ways of increasing the market for BC coal. We deal with each in turn, including the views of TISC and the final view of government, as embodied in the White Paper of March 1993.

On the question of the economic life of the Magnox nuclear stations, TISC took the same view as BC, namely that the matter ought properly to be judged, not on the 'accounting' costs of these stations (which included unavoidable past construction costs and future decommissioning costs) but on their avoidable costs. However, there were considerable difficulties in determining avoidable costs. While most operating and maintenance costs were avoidable, Nuclear Electric (NE) regarded the greater part of nuclear fuel costs (fuel supply, reprocessing and waste disposal) as unavoidable, particularly as NE had a 'take-or-pay' arrangement with British Nuclear Fuels (BNFL) for the reprocessing and waste management of Magnox spent fuel. Although, as TISC observed, 'a commercial commitment to future fuel services does not itself mean that the costs are unavoidable in national resource terms' (HC 237, para. 114), BNFL argued that its own costs were also largely unavoidable. Further, a rapid closure of the Magnox stations might well increase the costs of dealing with a greater quality of Magnox fuel requiring reprocessing in the short term (HC 237, para..116). In the White Paper, the government threw no light on the BNFL element in the Magnox costs, but stated that 'the range of 1.3 to 1.5p/kWh as the avoidable cost of additional output from the Magnox stations compares favourably with the avoidable costs of generation from existing coal-fired stations' (Cm. 2235, para. 7.78); and that 'the Government concludes from this analysis that there is no economic justification for requiring Nuclear Electric to close any of its Magnox stations before the end of their planned lifetimes' (para. 7.80). On the case of Sizewell B, the PWR nuclear station due to be commissioned in 1994, TISC noted that NE had already spent some 85 per cent of the total construction costs and that its fuel and operation costs were expected to be lower than for Magnox or AGR nuclear stations. Total avoidable costs (including the cost of completing the station) were expected to be about 1.4p/kWh, and that there was no economic case to suspend construction or mothball the station (HC 237, para. 119). The government White Paper did not mention this possibility (and by implication, rejected it).

The government also rejected the view put forward by some parties, both to the Coal Review and to TISC, that the Fossil Fuel Levy (FFL) income received by Nuclear Electric constituted an operating subsidy to nuclear power; and noted that the TISC proposal that NE should cease to receive this income would have no effect on the market for coal. FFL income was designed to allow NE to meet its future unavoidable decommissioning liabilities, and 'to the extent that Nuclear Electric is unable to meet its liabilities because of a reduction in its levy income, the taxpayer would have to meet them instead' (Cm. 2235, para. 16.5). Indeed, the government had a strong interest in maintaining the cash flow from nuclear stations for as long as possible, thereby postponing the inevitable decommissioning costs.

So far as 'committed' CCGTs were concerned (i.e. those gas-fired stations already operating, under construction or financially committed), BC recognised that even if measures were taken to make 'pass-through' of costs to RECs more difficult, 'the likely impact would be principally on the RECs' profits rather than in discontinuation of the projects or reduction of gas burn' (para. 3.20). TISC also noted that 'once a CCGT is built, the capital has been sunk, and the appropriate comparison is then between the avoidable costs of *existing* gas-fired stations and existing coal-fired stations, so it is not certain that coal-fired stations would benefit' (HC 237, para. 243). And, although 'their new gas-fired stations undoubtedly have a protected market position by virtue of the IPPs' contracts with RECs, nevertheless given that the IPPs entered the electricity market in good faith and with government encouragement, we would not support any action which penalised them for having done so' (HC 237, para. 250). The government White Paper also came out clearly against intervention to reduce the level of gas use by CCGTs which had commercially-negotiated contracts (Cm. 2235, para. 10.38).

The White Paper also considered the question whether the RECs had met their 'economic purchasing obligation' when contracting to buy CCGT-based electricity, which had been subject to two reports (in December 1992 and February 1993) by the DGES (Professor Littlechild, the electricity regulator). The White Paper pointed out that the DGES was considering, not the *cost* of generating electricity, but the *price* at which the generator offers that electricity to potential buyers (in this case, the RECs). The DGES had found that the IPP contracts appeared to compare well with other contracts available in terms of price, and in addition gave the RECs a greater diversity of fuel source and supplier and reduced vulnerability to environmental regulation. Accordingly, Professor Littlechild found no basis for concluding that the RECs had breached their economic purchasing conditions, although he did however record that he was not convinced that the *costs* associated with producing electricity from coal-fired plant were sufficient to justify contract prices as high as those offered to the RECs by National Power and PowerGen. (Cm. 2235, 10.43–45). We have noted earlier (Chapter 4) that the 'dash for gas' was facilitated by the Regulator's approach of looking at relative *prices* rather than costs, and by his emphasis on the development of competition by encouraging IPP generation, using CCGTs. The government clearly endorsed this approach, which had worked to the disadvantage of coal. Moreover, the White Paper offered no mitigation of the effect on the coal industry of the 11GW of gas-fired plant already committed, which would in due course, take over 30m. tonnes of BC's sales in the power station sector.

So far as new plants were concerned, BC had advocated, in its submission to the Coal Review, that the government should cease giving further Section 36 Consents for CCGTs. They said that 'the expansion of gas-fired power stations is of concern not only because the long-term contracts on which they are secured are likely to lead to the closure of potentially economic coal mines, but also because it is doubtful (and certainly has yet to be demonstrated) whether all of these stations, the bulk of which are not yet constructed, will produce electricity more cheaply than the existing coal-fired power stations they are replacing' (para. 3.19). TISC, while not going so far as BC, thought that 'there should be very careful consideration before any new applications for section 36 consent for CCGTs are granted' (HC 237, para. 246), on the grounds that 'at present gas prices additional CCGTs are unlikely to be cheaper than existing baseload coal-fired stations without FGD, and very unlikely to be substantially cheaper'; further CCGTs were not then needed to ensure compliance with European SO_2 emission limits under the LCPD; 'we do not regard IPPs as contributing significantly to genuine competition', and 'many of the largest projects for additional CCGTs are promoted by the main generators rather than IPPs' (HC 237, para. 245). The government's reaction to this question in the White Paper was equivocal. It admitted that 'at prices in the prospective contract for British Coal supplies from April 1993 onwards, the avoidable cost of generating electricity from large existing coal-fired stations (using 500MW turbines) is generally less than the full costs of a new CCGT, assuming that both run on baseload', the comparison could be affected by changes in gas and coal prices, and in particular future environmental constraints (Cm. 2235, paras. 7.50–7.54). But this was beside the point, since the government reaffirmed its previous policy on Section 36 consents, namely, that 'as a general rule, matters such as the need for a generating station, its capacity, choice of fuel used and type of plant are commercial matters for the applicant' (Cm. 2235, para. 13.34). The government saw 'no need to depart from its previously announced policy in relation to new power station consents' (para. 13.35). On this issue, despite the fact that the continuing pressure to build new CCGTs arose in large measure from the imperfections of competition in generation in England and Wales, the government were prepared to disregard the cumulatively adverse effects on the market for coal.

BC also sought to eliminate the loss of 6m. tonnes of coal sales arising from the imports of electricity via the 2GW transmission link with France. Originally, this had been intended to serve as a means of load balancing between the UK and French systems, but in fact, the link had operated almost at baseload, importing surplus French (nuclear) output. There were two further causes for complaint. First, nuclear electricity imported from France was 'non-leviable' (i.e. not subject to the 'Fossil Fuel Levy') and was therefore able to be sold at a higher price to RECs – in effect giving Electricité de France (EdF) a subsidy. Second, although EdF thereby had entry to the UK market, the potential to export (coal-based) UK electricity to France was in principle very limited, as a competitive wholesale market, equivalent to that which existed through the electricity Pool in England and Wales, did not exist in France (para. 3.24). TISC agreed, and recommended that 'electricity supplied from France cease to be non-leviable, and that EdF's ability to negotiate contracts to supply baseload electricity from 1993 be made conditional on UK generators having non-discriminatory access to the French electricity market and through the French transmission network to other countries'. TISC believed that such measures would ensure equal trade both ways through the interconnector, increasing the UK coal market by 6.5m. tonnes (HC 237, para. 142). However, the government took the view, based on legal advice, that any measures it might take to present or restrict trade through the interconnector world breach Article 30 of the EEC Treaty and could result in very substantial damages (Cm. 2235, para 7.102), and that EdF's 'non-leviable' stations could not be removed without giving EdF the benefit of levy payments, thus reinforcing their incentive to export to the UK (para. 7.104). In effect, the government concluded that there was nothing that could be done on this point to increase the UK coal market.

All the proposals by which BC had sought to expand overall UK coal *consumption*, were rejected in the government White Paper as infeasible or undesirable. Thus far, the government gave no indication as to how the market for coal might be increased during the following five years; and there were no proposals by TISC, not considered by the government, which would have had such an effect.

Yet BC's alternative scenario, set out in its submission to the Coal Review, whereby an additional 15m. tonnes per annum would be sold to the ESI in England and Wales over the following five years, depended overwhelmingly upon a larger UK market for coal, from whatever source, by curtailing the use of gas-fired plant, nuclear energy, and electricity imports through the French interconnector (See Table 5.2).

In addition, BC sought to safeguard its share of the total UK coal market by limiting the extent to which the generators would use coal stocks instead of BC current output, particularly in 1993/4 and 1994/5. As BC pointed out in their submission to the Coal Review, the generators were required by government to maintain a minimum level of coal stocks to provide for security of supply to electricity consumers, and that limit was intended to be reduced to 10m. tonnes by April 1993. As the generators at that time (end 1992) had some 33m. tonnes of stock, there was a real danger of a very large stock lift (BC assumed 22m. tonnes over two years). Indeed, this was a major component in

Table 5.2: BC Alternative Projections in Coal Review Submission Coal Consumption by ESI (England and Wales). Million Tonnes

Supplied by	1993/4		1994/5		1995/6		1996/7		1997/8	
	A	B	A	B	A	B	A	B	A	B
British Coal	40	55	30	45	30	45	30	45	30	45
Imports	9	8	10	10	11	8	8	8	8	9
Other UK producers	4	4	5	5	5	5	5	6	6	7
Generators' stocks	11	6	11	7	0	2	0	1	0	0
Total	64	73	56	67	46	60	43	60	44	61

A = 'Unchanged policies'
B = Including measures proposed by BC

Note: Derived from BC Submission to Coal Review Tables 1, 3 and 6(c) - requiring marginal adjustments to import figures regarded as residual

the arithmetic which had led to the 'cliff edge coal crisis' of October 1992. BC therefore advocated retaining a higher minimum strategic stocking requirement than intended, and spreading the stock reduction over a longer period. TISC had some sympathy with these views, and recommended that the government should require the generators to hold total stocks of not less than 20m. tonnes of coal (HC 237, para. 259). The government's response to this point in the White Paper noted that the closure announcements made in October 1992 assumed that the generators would reduce their stocks at the rate of 11m. tonnes a year in 1993/4 and 1994/5 (Cm. 2235, para. 13.20). The government's powers to set minimum stock levels were constrained by its duties under the 1989 Electricity Act, 'and it would not be lawful to use them simply to enhance the market for British coal. Furthermore, there would be no justification for taking new powers whose sole purpose would be, in effect, to force private sector companies to buy goods which they did not want to buy, for the benefit of another commercial entity' (Cm. 2235, para. 13.22). The government therefore restricted itself to saying that it would be taking forward its consultations with the generators about stocking arrangements for 1993/4 'as a matter of urgency' (para. 13.23) but without any indication of the likely outcome. The position the Major government took on power station stocks was in marked contrast to that taken by Mrs Thatcher, who had regarded the continuation of high stock levels as a necessary safeguard against the resurgence of major industrial disputes in the mining industry. By implication, by 1993 the government no longer took this threat seriously.

In the absence of government measures to increase the market for coal in the UK, and with little expectation that the rundown of generators' coal stocks over the next two years would be significantly reduced, the only way in which BC sales could be increased on a sufficient scale to avoid many of the pit closures was by reducing imports. But, as the White Paper was to point out (Cm. 2235, paras. 11.38–40), of the 20m. tonnes of coal imports in 1992, some 12m. tonnes were of prime coking coal and other grades of which the UK had insufficient sources of supply. Thus, any scope for displacing imports would be largely confined to steam coal imported by the electricity generators, then running at about 8m. tonnes a year.

Although BC had submitted to the European Commission in 1990 (updated in 1992) an 'anti-dumping' complaint against coal imports from certain countries into the European Community, and had unsuccessfully urged the government to encourage an investigation, there was little real hope of success. The grounds for complaint were that some countries sold at lower prices on the export market than to the domestic market; and at prices below the costs of production. (BC did both of these things). But the dumping allegations related to only 6 per cent of the total UK coal market, and any restriction of imports from the countries in question would have easily been made up from other international suppliers. In its submission to the Coal Review, BC did not propose any measures to restrict steam coal imports. Rather, it sought to establish that BC output would compete with imports at *inland* power stations by 1997/8, given planned reductions in operating costs. 'For burn at coastal power stations, however, we consider that it will be much more difficult to compete profitably. In the absence of contracts with the generators at the levels currently envisaged, imports to coastal stations would be higher, even in 1997/8' (para 3.39).

The White Paper noted that the average price of internationally traded steam coal in the year before the October 1992 closure announcement had been about £1/GJ delivered in large vessels into North West Europe (Cm. 2235, Table 7.2). Although the evidence to TISC, and the government's own advisers (Caminus) had pointed to a range of future prices (depending both on the $ price and the £/$ exchange rate) the White Paper made a reasonable judgement that the evidence put forward pointed to only modest price rises, or none at all, in the years ahead (Cm. 2235, paras. 7.40 and 7.41) although with high uncertainty in the longer term. It is not unreasonable therefore to make a 'central' judgement as to how things looked at the time by examining the pithead costs required to be competitive on the basis of an ARA price of £1/GJ equivalent, and using the UK transport costs quoted by TISC (HC 237, Table 7, para. 41), which were similar to those quoted by the White Paper (Cm. 2235, Table 7.2, Footnote 5). The results are shown in Table 5.3.

Table 5.3: Pithead Costs Required for Competitiveness with £1/GJ ARA Price. £/GJ, 1992 Money Values

		Inland Power Stations	Coastal Power Stations
	ARA Price	1.00	1.00
Plus	import transport cost	0.35	0.22
Less	BC transport costs	(0.17)	(0.43)
	BC's 'Competitive' pithead price	1.18	0.79

Against this, BC projected further reductions in operating costs to show that by 1997/8, on average, BC deep-mined output would be competitive with imports at inland power stations, which accounted for the greater part of UK coal use by the ESI (See Table 5.4).

Table 5.4: BC Deep-mined Output, Costs and Productivity. 'Current Strategic Plans'

	1991/2	1993/4	1997/8
Deep-mined output (m.t.)	70.6	32.2	27.1
Average deep-mined cash operating cost (£/GJ 1992/3 money)*	1.73	1.40	1.25
O.M.S. (tonnes)	5.31	8.17	9.37

Source: BC Submission to Coal Review : Table 4

* Excluding changes in working practices requiring legislation, estimated to reduce average costs by 15p/GJ by 1997/8.

The cost reductions would be achieved by continuing BC practices of improvements in mining techniques and equipment, and reductions in manpower. In addition, BC claimed that further cost savings were achievable given changes in working practices to establish the optimum shift length and supervisory management structure. The major difficulty arose from the Coal Mines Regulation Act 1908, which stipulated that mineworkers may not remain below ground for more than $7^{1}/_{2}$ hours, plus one 'winding time' in any 24-hour period. This became an onerous condition as BC pits aged and coal faces moved further away from the pit bottom. The 1908 Act effectively prevented BC from negotiating new contractual arrangements with the Mining Unions to secure more productive use of working time – for example a 36-hour week with four nine-hour days. The government accepted this argument. It noted (Cm. 2235, para. 12.30) that the Coal Industry Act 1992 already provided for the 1908 Act to cease to have effect on a date appointed by the Secretary of State by statutory instrument, and that the government intended, subject to a consultation process, to remove the impediment to efficiency represented by the 1908 Act.

Given this change by government, the BC view was that by 1997/8, its *average* deep-mined operating cost would be about £1.10/GJ (1992/3 money): so that, on average, deep-mined coal was likely to be profitable at import-related prices at inland power stations. This view was endorsed by the government and their advisors: 'The Boyd and Caminus reports broadly confirm that, if more productive mining techniques and working practices are implemented and the mining community adopts a proactive business attitude, British Coal can be competitive with imported coal at inland power stations. On this point, they are supported by other evidence quoted by TISC' (Cm. 2235, para. 12.25).

On the other hand, there was nothing seriously to challenge BC's own acknowledgement that it had little chance of competing at coastal power stations removed from the coalfields. (This argument applied with even more force to steam coal exports, as was acknowledged in the White Paper, Cm. 2235, para. 11.34). Given that most, if not all, projected coal supplies to inland power stations were to be secured by the contract volumes already proposed, on what basis therefore, could BC's total sales to power stations be increased at the expense of imports? The economic case for increasing BC sales in this way was very weak. As TISC had pointed out (HC 237, para. 208) BC's required competitive selling price would fall, and the cost of displacing imports would increase as the tonnage rose, since BC then had increasingly to displace imports at power stations where its own transport costs were least favourable. Moreover, the marginal BC *supply* would properly be represented by collieries with above-average costs: indeed a number of collieries had costs unlikely to be competitive even at inland power stations (Cm. 2235, Table 12.3). Thus, the displacement of imports at the margin would involve a large and continuing cross-subsidy.

The government was against any *direct* action to curtail imports, since there would be a serious risk that such action would be incompatible with the UK's EC and GATT obligations (Cm. 2235, para. 11.41). But the White Paper went further:

The Government sees no case for imposing restrictions on the import of coal, even if this were possible within its EC and other international treaty obligations. The Government is a strong advocate of free trade policies and has been in the forefront of efforts to bring the Uruguay GATT round to a successful conclusion. The existence of alternative sources of supply is important to ensuring the security of the United Kingdom's electricity supplies, and the need to compete with coal produced elsewhere is an important spur to further improvements in the British coal industry's own performance (Cm. 2235, 11.43).

BC had argued in their submission to the Coal Review, that its share of the UK market could be enhanced if the RECs' franchise limit should be maintained at 1MW (rather than being reduced to 100kW in April 1994) and extended beyond March 1998, since this would facilitate the pass-through of higher BC coal costs over a greater tonnage and for a longer time, through the 'back-to-back' contract mechanism. In principle, this approach could have been used to displace imports without financial penalty to the generators, with the extra cost being borne by domestic electricity consumers. However, the government rejected this idea. As it said in the White Paper (Cm. 2235, para. 10.30): 'The Government believes that the electricity market needs more competition, not less.' It 'wishes the movement towards greater competition to continue, and accordingly does not propose to make any change in the present arrangements providing for the reduction of the franchise limit to 100 kW from 1994 and its complete abolition from 1998'.

The Political Settlement of March 1993

By March 1993, therefore, the position was that the government had rejected all measures proposed by BC and others to increase the UK market for coal over the following five years; it had rejected significant mitigation of the problem of the huge 'overhang' of generators' coal stocks (other than agreeing to talk to the generators about planned stocks in 1993/4); and it had rejected ways of increasing BC's share of the UK coal market, using extensions of the RECs' franchise markets or *direct* limitation of coal imports. The government was also against specific measures to reduce opencast production in order to create a bigger market for deep-mined coal. Although environmental planning guidance on new opencast sites was subject to review (Cm. 2235, para. 13.31), it was a matter for BC to decide the level of opencast output for which it wished to aim (Cm. 2235, para. 13.33) (although the government noted that BC expected its opencast production to decline somewhat).

But, if the government's position was that little or nothing could or should be done to ameliorate the market situation which had led to the 'coal crisis' in October 1992 when the 31 closures were announced, what did it propose to do, given the urgent need to find a politically acceptable solution?

The government was heavily conditioned by the TISC report, which had been published on 26 January 1993, and which gave the strong impression that, provided the political will existed, much could be done. Specifically, TISC recommended that the government should provide a subsidy to the generators in England and Wales to burn up to 16m. tonnes of deep-mined BC coal per annum above the quantities of 40m. tonnes falling to 30m. tonnes, which they were already expected to contract for in 1993–8 (thereby effectively halving the projected fall in sales). The subsidy should be equal to the difference between the delivered costs of the additional BC coal and imported coal. TISC also recommended that the government consider financial assistance to contracts undertaken in the non-ESI market for up to 3m. tonnes per annum for five years. The subsidy would be conditional on BC meeting cost reduction targets and keeping open as many pits as possible. The total cost of the subsidy would be unlikely to exceed £500m. over the five years (HC 237, paras. 228, 229 and 231).

Although BC expressed great scepticism on the feasibility of the additional market likely to arise from the TISC proposals (BC response to TISC report, 11 February 1993, para 3.18), the *political* significance of the report was that it set a bench-mark of acceptability, in particular for Conservative rank-and-file opinion. In other words, a settlement had to be produced which could be claimed to have saved about half the threatened pits. For this reason, and because all other routes to additional BC sales had been rejected, the government reacted favourably to the TISC suggestion on subsidy. Heseltine later admitted that Richard Caborn, the Labour chairman of TISC, inadvertently 'threw the Government a lifeline' (Heseltine p.442). As it said in the White Paper: 'The Government accepts the Committee's recommendation that a subsidy should be available for additional sales by British Coal to the England and Wales generators and the proposals it has put forward in negotiations between the generators and British Coal incorporate several of the suggestions on the form the subsidy should take' (Cm. 2235, para 16.14).

It needs to be emphasised that the conclusion of the Coal Review was not an exercise in industrial management but a *political* settlement, designed to secure a favourable vote in the House of Commons (particularly by reassuring the Conservative Party), and to avoid a negative reaction from public opinion. This was the purpose of Michael Heseltine's statement to the House on 25 March 1993, on the day that the White Paper was issued. There were two main approaches to secure this end: firstly, to show that there was a credible way by which the threat of imminent and permanent closure could be removed from a majority of the 31 collieries announced for closure on 13 October 1992; and secondly to facilitate the privatisation of British Coal. Indeed, the government saw those approaches as linked, particularly as it was a widespread view in the Conservative Party (and, indeed, in the government itself) that much of the problem of closures had been brought about by incompetent management by the nationalised British Coal. As the President's statement to the House stated:

> British Coal's prospects of winning and holding a share of the wider market for coal depend on increased competitiveness. There is a critical role here for the private sector. We believe that privatisation is the only way of enabling this industry to take full advantage of the opportunities the market offers. British Coal will therefore begin immediately to prepare for full privatisation. They have confirmed that in advance of full privatisation any pits which the Corporation does not itself wish to keep in operation will be offered to the private sector (Statement para 22).

On the issue of the market, the President began by informing the House that the electricity generators (in England and Wales) and RECs had now formally confirmed their intention to enter into base contracts and the necessary back-to-back contracts which would enable BC to sell 40m. tonnes in the first year and 30m. tonnes for each of the following four years. (Statement para 5) But as these were the tonnages which had given rise to the October 1992 closure announcement which had triggered the 'coal crisis', the government needed to make proposals which would result in additional sales. The President's statement explained why the government had decided to take no action to curtail the use of gas, nuclear power, or electricity imports by the French interconnector, and drew attention to the high level of generators' coal stocks. The only market measure brought forward was a subsidy to supplement power station coal imports:

> I have authorised British Coal to negotiate for future contracts on the basis that they would supply at a world market related price. The Government is prepared to subsidise the difference between this price and British Coal's cost of production. This subsidy will apply for any additional tonnages that British Coal is able to sell to the generators, whether or not these are on long-term contracts. The subsidy will reduce progressively over the period

to full privatisation. The amount of subsidy will depend on the outcome of commercial negotiation but we are prepared to embrace the range of figures put forward by the Select Committee (Statement para 26).

But what did this proposal amount to? Even if the generators were to agree to cease importing any coal and to substitute these supplies with subsidised BC coal, the increase in BC sales would be no more than about 9m. tonnes per annum. Such an outcome was unrealistic, given that the generators wished to diversify their coal supplies (and would therefore wish to retain some imports), and the lack of any real incentive to switch to subsidised BC coal if these substitute supplies were to be at the same price as imports. Indeed, in his statement to the House, Michael Heseltine added that 'I cannot guarantee that supplementary sales will be achieved by British Coal. But both generators have said that they will continue negotiations for such sales' (Statement para 28). Moreover, Heseltine made his announcement even though, as he said in his memoirs, 'I knew that the generators would not in practice take up any significant extra tonnage' (Heseltine p.442).

Nevertheless, in spite of the nebulous character of the sole specific government proposal to increase BC sales, and the great uncertainty over generators' plans to use their coal stocks, the President was able to tell the House that British Coal, having been given an advance copy of the White Paper, had that day made an announcement substantially reducing the number of immediate closures and that 'the future of coaling in individual pits will depend on the extent to which intensive efforts over the coming months identify a market for their product or realistic prospect of sale to the private sector' (Statement paras. 30 and 31).

The content and wording of the British Coal statement of 25 March, timed to coincide with Michael Heseltine's statement to the House of Commons, had been the subject of intense discussion between BC and Ministers for several weeks before the issue of the White Paper. The President was clearly very concerned that British Coal should be seen to be supporting the government in the political process by welcoming the White Paper, and by ensuring that the public statements of the Corporation were entirely in tune with those of the government; since otherwise the government's ability to win the vote in the House might be put in jeopardy. This placed the Corporation in some difficulty, since in practice it had no alternative but to do what Ministers wanted. On the other hand, BC had two concerns in particular, which were communicated to Ministers. First, there was anxiety that if the number of pits to be announced as continuing in production was too high, this might make it difficult to rebut challenges that the exercise to find additional markets had no real substance; second, although BC confirmed its willingness to offer any pit proposed for closure by BC for sale to the private sector, there was a danger that any such pits might seek to secure access to the 'core' contracts with the generators and/or the government's proposed subsidies, which would be counter-productive so far as British Coal was concerned.

But the die was cast. While welcoming the White Paper, the BC statement said that 'it was clear that despite the offer of price support it would not be possible to secure a market for all of the 21 collieries covered by the Review. However, the Corporation felt the circumstances merited maintaining 12 of those collieries in full production while it carried out the market testing which would determine the volume of extra sales which could be secured.' Of the remainder, BC proposed ceasing production at six collieries which would remain in 'care and maintenance' while the potential additional market was explored. A further colliery would cease production but be redeveloped, and two were proposed for closure. BC also added that it saw nothing in the White Paper to alter its current proposals to close the ten collieries which were identified for closure before the Review (and which had been subject to the legal challenge of judicial review).

In addition, BC announced that the redundancy terms agreed with the government in early October 1992, which previously were due effectively to end in March 1993 would now continue to December 1993; and on this basis the BC gave the government assurances that it would not pre-empt the outcome of consultations about future closures and that redundancies would be voluntary (*BC Annual Report, 1992/3*, page 14). In this way, not only was more flexibility created on the phasing of closures, but also BC was able to resolve the legal challenge which the unions had made to the October 1992 closure announcements.

The total output capacity of the twelve 'reprieved' pits which were to continue in production was 13m. tonnes per annum: significantly more than the quantity of coal imports that realistically could be substituted. The arithmetic of BC's announcement was the result of political calculation rather than market expectation. Nevertheless, BC's proposals, as announced by the President to the House, together with the requirement that all collieries surplus to BC's requirements would be offered to the private sector before final closure, and the commitment to early BC privatisation, reassured the Conservative Party. In addition, the crucial support of the Ulster Unionists was secured by a promise to bring forward plans for an electricity interconnector between Northern Ireland and Great Britain (recalled by David Hunt, 13

December, 1996 on Radio 4), and to set up a new Select Committee for Northern Ireland affairs. All this was sufficient to provide the government with a majority of 22 in the House of Commons debate on the White Paper on 29 March. The Conservative backbench rebellion had been reduced to only four, with four abstentions. The government also gained from the passage of time. Once it had won its vote on 29 March, public interest, already less intense than in October 1992, continued to wane. This had been a political triumph of presentation over substance.

The Coal Review and Public Policy

Several aspects of the Coal Review are important in throwing light on the nature of the government's policy towards the coal industry within the wider context of public policies generally.

Firstly, although there had been much preoccupation with establishing politically acceptable phasing of closures and manpower rundown, this was not in any sense part of a policy of job protection in areas of high employment, still less because it was judged that mining employment had a particular value. Yet this had been one of the most intense areas of debate during the time of the Coal Review. The Employment (Select) Committee said in its report 'Employment Consequences of British Coal's Proposed Pit Closures' (HC 263 published 13 January 1993) that BC's closure proposals, including their indirect effects, could jeopardise the employment of up to 100,000 people with little hope of alternative jobs (HC 263, para. 20).

Accordingly, the Committee declared that

> The financial cost of resulting unemployment should have been calculated and should have been part of the equation when deciding the future of the pits. The Department of Employment should have a clear and important involvement in the decisions resulting in mass redundancies, especially by making prior forecasts of job losses and of their financial and social cost and the likelihood of those who lose their jobs finding other work. The true national costs should be taken fully into account especially where large-scale redundancies are to be created in a publicly owned industry such as coal (HC 263, para. 25).

And, after taking evidence from, amongst others, the Archbishop of York and Cardinal Hume, the Committee observed that 'the announcement of the sudden closures has had a devastating effect on the communities which depend on the pits. The social effects of pit closures should be taken fully into account when deciding whether or not to keep pits open' (HC 263, para. 66). Indeed, in their conclusions the Committee stated that 'no pit should be closed unless and until there has been full consultation and complete consideration of all the factors set out throughout this Report and until every possible effort is made to safeguard not only the miners concerned but other affected workers and their employers' (HC 263, para. 82).

This was a re-run of arguments which had been used by Select Committees on earlier occasions, both on the substance of the issues, and by way of criticism of the government and British Coal. The Committee noted that, in spite of the 'importance of the issues raised', the Department of Employment and the Secretary of State (Gillian Shephard) had had little, if any, role during the summer and autumn of 1992 on the future of the coal industry (HC 263, para. 20). Yet, in the politically charged circumstances of January 1993, the Committee (with its Conservative majority) was reluctant to publish a report directly criticising Mrs Shephard, and so attempted to place the blame on BC on the grounds that it had made no significant effort to arrange a 'proper briefing meeting' with the Secretary of State for Employment (HC 263, para. 17). This trivial charge, with the ludicrous implication that the failure to arrange such a meeting resulted in the Employment Department being unaware of important matters upon which they could have taken effective action, provoked British Coal into issuing a Press Release which directly criticised the Employment Committee. And on the main policy issue, the government brushed aside the Select Committee's case. In the White Paper, they pointed out that between 1987 and October 1992, when 84,000 jobs were lost in British Coal, 'total unemployment in the areas where pits have closed actually *fell*, and the fall was faster than in Great Britain as a whole' (Cm. 2235, para. 5.8); 'on a national basis four out of five people who became unemployed cease to be unemployed within a year .... Over time, as the economy adjusts and redundant workers find new employment, the loss of jobs from pit closures would be broadly matched by compensating developments elsewhere'. (Cm. 2235, para. 5.13, 5.14). In sum, the government was not prepared to use a calculus involving unemployment and social costs to slow down the rate of reduction in coal mining employment. Indeed, as we have noted before, the policy of continuing to finance BC generous redundancy and social costs was designed to *facilitate* the loss of mining jobs in a politically acceptable way. The only concession that the government was prepared to make was a decision to increase to £200m. the sums available for 'regeneration measures' in areas most likely to be affected by closures, in addition to the package of measures to assist coalfield communities, which the government had announced in October 1992 (Cm. 2235, para. 5.17).

Secondly, the March 1993 Coal Settlement further clarified the government's attitude to coal in relation to environmental policy. We have noted earlier that the government had been unwilling significantly to mitigate the impact on the coal industry of increasing environmental pressures. The White Paper reiterated this position. The government noted that TISC had acknowledged that 'burning coal results in greater emissions of SO_2, NO_x and CO_2 than burning almost any other fuel, and that on most estimates the economics of burning coal will therefore be more strongly affected by environmental regulation' (Cm. 2235, para. 8.5). There was no suggestion that the government would look favourably on suggestions to relax emission limits to assist coal: rather the opposite. The White Paper notes that the LCPD was due for renewal in 1994; that during 1993 negotiations would continue on a revised SO_2 protocol to the United Nations Economic Commission for Europe's Convention on Long-Range Transboundary Air Pollution Control; and that further regulation could arise from the powers of Her Majesty's Inspectorate of Pollution (HMIP) under the Environmental Protection Act to ensure that best available techniques not entailing excessive cost (BATNEEC) were used to minimise harmful releases to the environment. (Such 'techniques' could include the use of low sulphur imported coal, or building CCGTs, as well as technical abatement measures at coal-fired power stations). The White Paper concluded that although existing measures (6GW of FGD committed, with the possibility of a further 2GW) were sufficient to meet UK emission targets under LCPD until 2003, 'the possibility that tighter limits may be imposed in the future, however, introduces an element of uncertainty, and hence a perception of increased commercial risk, which weighs particularly against coal' (Cm. 2235, para. 8.14).

The White Paper also drew attention to the problem of Global Warming and the government's commitment to ensure that CO_2 emissions were no higher in 2000 than in 1990. This target 'will already constrain coal burn by the year 2000. Any tightening of the CO_2 target in future reviews which are scheduled before the end of the decade is bound to constrain further the amount of coal burnt unless very stringent measures to curb emissions are taken in other areas, including transport' (Cm. 2235, para. 8.28).

The TISC report had made a strong plea for government funding of 'Clean Coal Technology' (which the Committee had already investigated in 1991) which incorporated methods of sulphur capture and also, because of higher thermal efficiency, emitted less CO_2 per unit of electricity. The Committee noted that although considerable work had been done by BC's Coal Research Establishment (CRE) in the development of a new 'clean-coal' combined-cycle technology (known as the 'Topping Cycle'), the crucial problem was the lack of demonstration plants which was an essential step in progress towards commercial use (HC 237, para. 278). They declared that 'greater assistance towards clean coal technology would do more than anything else to demonstrate that the Government is committed to the long-term future of the British Coal industry. Such a commitment could best be shown by commitment towards the present Bilsthorpe proposal, other demonstration work connected with the topping cycle, and a demonstration coal gasification generating plant' (HC 237, para. 279). The Committee also recommended that uncertainty on the future funding of the CRE should be resolved (HC 237, para. 280). The government's response on clean-coal technology gave little encouragement to its advocates. 'To ensure that there is adequate support for R&D on coal utilisation in the United Kingdom', the government decided to increase its support for coal R&D from £3m. to around £7m. per annum over the following three years (i.e. an extra £12m. spread over three years). This was barely sufficient to sustain the level of activities at CRE, and did nothing to address the question of funding demonstration plants, on which the government's view was even more negative. It noted that two advanced systems, Integrated Gasification Combined Cycle (IGCC) and Pressurised Fluidised Bed Combustion (PFBC) were currently being demonstrated on a commercial scale in Europe, but that all such plants required special finance (Cm. 2235, para. 5.14), and that none of the proposed demonstration plants in the UK was currently economic and would proceed only on the basis of a substantial level of government funding (Cm. 2235, para.15.17). The government concluded that 'it would not be justified in granting substantial public money towards a clean coal demonstration plant at the present time', on the grounds that such a plant would 'not materially affect the number of coal mines kept in operation in the United Kingdom over this decade' (Cm. 2235, para. 15.18). Instead, the government suggested that BC's Topping Cycle R&D programme should now be taken forward by an industry-led consortium with a direct interest in the commercial development of the technology (Cm. 2235, para. 15.21). Whatever the merits of the government's position on UK clean coal technology (and the coal interest had tended for some time to exaggerate the case as a means of helping UK deep mines), the government's refusal to provide substantial public finance for demonstration plants (rather than R&D) was another symbolic blow to the coal industry.

The terms of reference of the Coal Review, as announced by Michael Heseltine on 26 October, said that the future of the pits then threatened by closure would be considered 'in the context of the Government's energy policy'. The description of that policy set out in the White Paper gave no comfort for those who believed that there should be an element of special treatment for coal on 'energy policy' grounds. The government stated that the aim of its energy policy was to 'ensure secure, diverse and sustainable supplies of energy in the form that people and businesses want, and at competitive prices' (Cm. 2235, para. 3.1). But, as so often with policy formulation, the ways and means are the real defining elements. In particular, the government set itself against intervention, since it believed that its aim 'will be achieved most fully through the mechanisms of the market. The market is the most effective and efficient means for meeting energy needs ... Government should not attempt to impose all-embracing plans about how much energy of what kind should be produced or consumed by whom' (Cm. 2235, para. 3.2). 'Security and diversity of supply are best achieved through the operation of competitive and open markets' (para. 3.3). Thus, 'the Government's role is to provide the [energy] sector, its customers and the public with a stable and effective framework of law and regulation, which will protect health, safety and the environment and allow competition to flourish. Government will carry out those duties which can only be performed by Government' (para. 3.15).

The government recognised that there were limits to the speed at which change to fully competitive market conditions could be accommodated (Cm. 2235, para. 3.21). But any concessions were to be time-limited. In the case of BC, the proposal for a subsidy for coal burn would 'reduce progressively over the period to privatisation, which the Government intends to achieve at the earliest practical opportunity' (para. 3.13).

The significance here of statements on the principles of energy policy is not to test their intellectual rigour. (For example, there is little discussion of how 'externalities' are to be dealt with or how potential conflicts between 'free markets' and environmental protection are to be resolved). Rather, the point is that these things are said as part of the government's White Paper on the Coal Review. If the general impact of market forces and environmental concerns was a contraction of the coal industry, there were no policy reasons why the government should stand in the way in the name of 'energy policy', nor (as we have already noted) to maintain employment levels in the coalfields.

The Coal Review process also effectively (at least so far as the government was concerned) disposed of two other arguments that had often been advanced in favour of 'special treatment' for coal, namely:

- Firstly, the huge size of UK deep-mine coal reserves ('at least 300 years'), in contrast to the 'limited' reserves of North Sea Oil and Gas, gave them great strategic value.
- Secondly, UK produces by far the cheapest coal in Europe and UK coal should therefore be safeguarded as part of a 'European' energy policy.

Firstly, the question of UK coal reserves. TISC drew attention to the fact that the '300 years' of reserves derived from earlier BC statements that the total amount of coal 'in place' less than 4000 ft. deep and in seams at least two feet thick, was 190bn. tonnes, of which 45bn. tonnes were 'technically recoverable', which was then divided by the 1974 'Plan for Coal' target output of 150m. tonnes a year. But this said nothing about what was *economically* recoverable. In evidence to the Committee, BC stated that capital investment in new mines looked very unlikely, so that the only coal reserves likely to be available at reasonable cost were those accessible from existing pits. British Coal estimated that the 50 existing pits (including those which had ceased coaling since October 1992) together with Asfordby new mine (then still under development) had 2186m. tonnes of 'reserves' of which 1102m. tonnes were 'reserves not able to support mining projects at present,' leaving 1084m. tonnes as likely to be economically recoverable (HC 237, paras. 161–163). John T. Boyd, the American mining consultants, whose report to the DTI in January 1993, gave individual pit details of UK coal reserves in the public domain for the first time, had similar overall results for 41 collieries (that is excluding the ten initially excluded from the Review). For the purpose of economic analysis, Boyds included all 'classified' reserves, but ignored 'un-classified' reserves requiring major capital investment to gain access, and took only 40 per cent of 'unclassified' reserves accessible by 'current and minor development', giving a total of just over 1 billion tonnes (see Table 5.5). In the White Paper the government pointed out not only that 'the vast bulk of the nation's coal resources are not within the take of currently operating pits' (Cm. 2235, para. 7.27), but also that 'even at existing mines it is not economic to recover all the coal that could be accessed' (para. 7.28) and that UK's coal reserves are small by international standards (para. 7.33). In answer to a question from TISC, BC had said that 'currently economically viable reserves

Table 5.5:   Reserves at 40 Existing Collieries and Asfordby: January 1993. Million Tonnes

(a) 'Classified' reserves	883
(b) 'Unclassified' : Current and Minor Development	384
(c) 'Unclassified' : Major Development	848
Total	2115
Total of (a) + 40% of (b) =	1037

Source: Boyd Report to DTI: January 1993

at existing mines would have a life of between 20 and 40 years, varying within that range according to the prices obtainable, and probably closer to 20 years than 40' (HC 237, para. 163). Moreover, because reserves are not distributed evenly over all collieries, an overall reserves/production ratio may mislead. It is notable that of the 2.1 billion tonnes of undiscounted reserves at existing pits quoted by Boyds in January 1993, almost half were at six collieries. This meant that, whatever the precise outcome of the Coal Review, British Coal's output was likely to decline rapidly in the early years of the next century as reserves were exhausted. The myth of '300 years' of UK coal reserves had been laid to rest for all those willing to study the published figures, and the government was able to make the most of the point.

Secondly, the question of the position of the UK coal industry within Europe. In December 1992, the European Commissioner for Energy, Cardoso e Cunha, told TISC that it would be a 'distortion' if the cheapest deep mines in Europe closed while heavily subsidised mines, particularly in Germany and Spain, continued in production (HC 237, para. 238). However, as the author can testify as a result of earlier discussions with the Commission in Brussels, there were no powers available to avoid this outcome, however sympathetic officials might have been. At the time of the Coal Review, there was some discussion about the Commission's informal rule (based on a Commission Working Paper in 1990) that state aids could apply to up to 20 per cent of the energy used to generate electricity, and whether this provided a mechanism for protecting UK coal (HC 237, para. 239). However, not only did this 'rule' have no legal force, it was regarded by the Commission in the case of the UK as being taken up by the support for Nuclear Power under the Fossil Fuel Levy, introduced as part of the ESI privatisation settlement of 1989/90. For the government, there were two issues: whether the limited subsidy proposal was compatible both with the current State Aids directive (2064/86/ECSC) and any subsequent replacement directive; and the clearance with the Commission of the new coal contracts between BC and the two main generators. Thus, the only real significance of the European Commission in this whole affair was as a regulatory body of last resort. Indeed, the very limited nature of EC involvement in the UK 'coal crisis' only served to highlight the impracticability of devising an effective protectionist Community coal policy, given the radically different positions taken by the UK and German governments on the issue of support for indigenous deep-mined coal, and the fact that the majority of member states had no significant coal industry, and were more interested in access to cheap coal imports from outside the Community. And, as to the argument that UK coal should be supported because it was the cheapest in Europe, the government was dismissive: 'The United Kingdom undoubtedly has the lowest cost coal industry in the Community. It depends, however, on its ability not to compete with other EC producers, but with other fuel sources and with coal imports from outside the EC' (Cm. 2235, para. 9.9).

By the end of March 1993, the government had extracted itself, with considerable political skill, from the largely self-inflicted crisis of October 1992, by means of a coal settlement which, at least to its own satisfaction and that of its supporters, appeared to have avoided most of the closures proposed in British Coal's draconian announcement of 13 October 1992, and to have done so in a manner wholly consistent with its 'free market' energy policy principles, most notably by announcing that British Coal would begin immediately to prepare for full privatisation and that the government would bring forward the necessary legislation as rapidly as possible. And so, some four-and-a-half years after Cecil Parkinson made his pledge to the Conservative Party Conference in October 1988, this 'ultimate privatisation' became the overriding political objective for coal.

CHAPTER 6
THE PRIVATISATION OF BRITISH COAL

British Coal's Tasks before Privatisation

For British Coal, privatisation had moved centre stage on the conclusion of the government's Coal Review. The preliminary work which had begun in June 1991, with the appointment of government advisers (led by N M Rothschild), had been suspended between October 1992 and March 1993 while the Coal Review was carried out, but would now resume. Paving legislation for privatisation (The British Coal and British Rail (Transfer Proposals) Act) had received Royal Assent in January 1993, and enabled BC to participate fully in the preparations for privatisation. Before the main legislation to privatise BC could be introduced, the Coal Review White Paper had signalled a number of interim measures, namely: to remove the limits placed on underground working hours by the Coal Mines Regulation Act 1908 (Cm. 2235, para. 12.30), (which change came into force in November 1993); and to introduce legislation to remove the restrictions in Section 36(2) of the Coal Industry Nationalisation Act 1946, under which BC was unable to license mines employing more than 150 workers underground (Cm. 2235, paras.14.31 and 14.32).

The removal of restrictions on BC's ability to license deep-mine operations was necessary to give effect to the government's insistence that in the interim period before full privatisation BC should offer to the private sector all collieries which were surplus to its requirements. This reflected the belief of the government and its supporters that private management would operate the collieries more efficiently and competitively, so that there was merit in introducing as much private management as possible in the period before full privatisation could be achieved, in order to test the validity of BC's proposed closures. As the White Paper said (para. 14.27), 'there are clearly considerable attractions to giving private sector mining companies the opportunity to run pits which would otherwise close. New management teams might be able to produce gains in efficiency which British Coal has not been able to secure.'

The government also made it clear that there was no intention of privatising BC in its existing form. Rather, its mining and other assets would be offered for sale in a number of separate packages. From March 1993, therefore, it was clear that British Coal's task was not only to secure its privatisation, but to facilitate its own abolition and the demise of its senior management. In its public rationale for BC privatisation, the government had stated in the White Paper: 'There are clear benefits to the coal industry and to consumers of coal in freeing the managers of Britain's coal industry from the constraints of public sector ownership. It is to a large extent these constraints that have brought British Coal to its present position, and there can be no confidence that within them the industry will achieve the dramatic changes necessary if it is to have a long term future' (Cm. 2235 para. 14.21).

Yet this was a very unbalanced view of the nature of the problems facing the UK coal industry, and British Coal in particular, from March 1993. If the sale of BC to the private sector was to be characterised as a political success several (related) conditions had to be satisfied: BC's mining business had to be sufficiently attractive to prospective purchasers (making competing bids under a 'trade sale') while at the same time yielding sufficient revenue for the Exchequer to avoid the accusation that the industry was being 'given away'; and the newly-privatised industry would have to appear to have a prosperous future ahead of it. In turn, this meant that a number of important tasks of economic management of the industry had to be largely completed by BC *before* privatisation. Firstly, notwithstanding the prospect of a large and rapid fall in demand, a balance between output and demand, capable of being sustained for a period after privatisation, had to be achieved: otherwise, potential buyers would be deterred; and it was also very much in the government's political interest to avoid a situation where large-scale closures immediately followed privatisation. Secondly, it was important that the programme of redundancies and manpower rundown, arising both from the reduction in capacity and improvements in productivity, should be largely complete by the time of privatisation; not only because the government would not wish to see large numbers of redundancies appearing to be the immediate and direct result of privatisation, but also because the costs of such redundancy (particularly if the very generous BC redundancy terms were carried over for a period) would be a major disincentive to potential buyers. Thirdly, the operating costs of continuing collieries had to be reduced to a point which would enable potential buyers to make significant profits immediately after privatisation. To complete these inter-locking tasks between March 1993 and December 1994 (when the sale of BC's mining assets was concluded) was a major undertaking. Because of the risk aversion of prospective purchasers and, especially, their financial backers, the actions required to ensure the successful privatisation of BC increased the pressure to reduce both output and manpower, and to reduce market uncertainty.

One essential element was already in place. We have already noted that the President of the Board of Trade's statement to the House on 25 March 1993 had announced that 'the generators (i.e. National Power and PowerGen) and the Regional Electricity Companies have now formally confirmed their intention to enter into base contracts and the necessary back-to-back contracts' which would enable British Coal to supply 40m. tonnes in 1993/4, and 30m. tonnes in each of the next four years. The prices were to be those included in BC's 'final' offer in September 1992: namely, pithead prices of £1.51/GJ in 1993/4, falling progressively to £1.33/GJ in 1997/8 (at September 1992 money values). These terms were agreed between BC and the generators at the same time as the White Paper was published (that is, only a week before they were due to come into force) and full contract documents were signed in June 1993. Also, in 1993, Scottish Power expanded their contract with BC to some 3m. tonnes a year to March 1998. These contracts with electricity generators, together with a small number of industrial contracts, meant that at least three-quarters of BC's likely sales over the following five years were secured on a take-or-pay basis, and at predictable prices, which were significantly higher than free commercial negotiations would have produced. Moreover, prospective purchasers of BC's mining assets would enjoy the benefits of these contracts for over three years after privatisation. The government-brokered contracts with the generators not only provided a clear commercial framework, but also played a vital part in reducing the considerable risks inherent in the whole process of coal privatisation. Indeed, without these contracts, privatisation could not have been carried through, except as a chaotic distress sale.

Notwithstanding the agreement of the contracts with the generators, BC's task of achieving a sustainable balance between supply and demand was complicated by the unsatisfactory outcome of the 'Coal Settlement' in March 1993, which (as we have seen) provided little reason to believe that significant sales additional to those specified in the new contracts could be secured, or that large-scale closure of deep-mine capacity could be avoided, particularly as the generators had made it clear to government that they were not prepared to curtail the planned rundown of power station coal stocks over the following two years. Nevertheless, on 21 May 1993, BC made formal offers to

National Power and PowerGen for the supply of additional coal. National Power indicated that they might be prepared to take more coal late in 1993/4, but PowerGen were more discouraging (*BC Annual Report 1993/4*, p.8); and it soon became clear that there was no immediate prospect of additional subsidised sales in the way envisaged in the White Paper. The position was made more difficult by the qualifying rules formulated by DTI, under which the subsidy announced in the White Paper would be obtainable. These were that the subsidy applied only to deep mines (whether BC or private sector); the sales were to be for electricity generation; and, crucially, the subsidised sales had to represent a genuine 'additional' market. Moreover, any mine saved from closure as a result had to have the clear prospect of being commercially viable without subsidy by 1995. Given the inherent difficulty of demonstrating that these conditions had been satisfied, they effectively ruled out subsidy for any British Coal supplementary sales to National Power and PowerGen. The strong feeling at BC was that the rules were so designed as to prevent the payment of other than a token subsidy.

It will be recalled that the President of the Board of Trade in his statement to the House on 25 March 1993 had said that 'the amount of subsidy will depend on the outcome of commercial negotiation *but we are prepared to embrace the range of figures put forward by the Select Committee*' (Statement para 26: author's italics). TISC had talked (HC 237, para. 209) of figures 'not exceeding' £200m. in 1993/4, or £500m. over the five years to 1998. In fact, in April 1993 agreement with the Treasury was secured for an initial provision of £120m. for the coal subsidy for 1993/4 (of which £100m. would be for BC and £20m. for the private sector). However, when putting forward this proposal for Parliamentary approval, the government wished to avoid criticism that this provision was far too low relative to the TISC figures. Thus it was stated that it was the government's intention to go back to Parliament with a new resolution if that proved necessary. Nevertheless, it became increasingly clear that the government had no intention of allowing large subsidies to be paid. Indeed, discussions between BC and DTI officials showed that formal applications for subsidy arising from offers to National Power were to be discouraged to avoid the potential embarrassments arising from rejection. In the event, the entire subsidy made available to British Coal was only £1.7m., relating to sales of 213,000 tonnes of coal from Ellington colliery in Northumberland to the Lynmouth Alcan smelter and to German and Danish power stations. No subsidy was ever paid to BC in respect of additional sales to National Power and PowerGen. The only specific proposal in the White Paper which was capable of providing additional sales to BC, to avoid colliery closures, and which had been central to the political settlement of the 'Coal Crisis' of October 1992, was shown to be without any real substance, so far as BC was concerned. And although the private sector obtained some £15m. of subsidy over a period of two years, this supported less than 1m. tonnes of additional sales.

This left BC with the task of pursuing the additional sales necessary to sustain those collieries 'reprieved' as a result of the Coal Settlement of March 1993, but without an effective instrument of subsidy. In the event, only 0.4m. tonnes of supplementary sales to the generators in 1993/4 could be negotiated. By the summer of 1993, BC stocks had reached 14m. tonnes and were increasing by 200,000 a week as a result of the (predictable) failure to secure additional sales, and the continuation of the output from pits 'reprieved' by the announcement of 25 March 1993. Moreover, although BC was able subsequently to agree with National Power for supplementary sales of some 6m. tonnes, these were to be spread over a two and a half year period from April 1994, and provided little offset to the further fall in the 'core' contract tonnages of 10m. tonnes per annum from April 1994. Thus, only six months after the government's Coal Settlement, it was clear to BC that there was a need for large output reductions if an unsustainable imbalance between supply and demand was to be avoided, and the progress towards 'successful' privatisation was to be maintained. Despite some pressure from the DTI to mitigate the *number* of closures by restricting the output of continuing collieries, BC rejected such a policy since it would directly conflict with the objective of bringing down average production costs per tonne as rapidly as possible. A large output reduction had to be achieved primarily by closing the least economic deep mines and, to a much lesser extent, reducing opencast by seeking fewer new site approvals. By September 1993 BC had concluded that as many as fifteen additional closures would be required by the end of March 1994. There followed intensive internal debate as to which collieries should cease production, in order to ensure that the collieries which survived were those with the best prospects after privatisation.

Of the 51 BC collieries in production or under development at the time of the 'Coal Crisis' of October 1992, ten had ceased operations by the end of March 1993, and by March 1994 a further 22 pits had stopped producing coal. This left just nineteen BC collieries producing or under development. However, these numbers alone are insufficient for an understanding as to what happened to the 31 proposed for closure in October 1992.

The analysis is complicated by the changing status of a number of collieries, some of which ceased production for a time (while the basic pit infrastructure was being maintained), and by the inauguration of the policy of offering surplus collieries to the private sector under lease/licence. Indeed, this complexity of analysis was politically convenient in that the underlying trends were more difficult to discern publicly. In addition, during the 1993/4 financial year, BC had to consider the cessation of production at some collieries which in October 1992 had been thought to have a secure future. The changing status of BC's 51 collieries, 31 of which were planned to cease production in the BC statement on 13 October 1992 is shown in Table 6.1.

Thus, in May 1994, thirteen of the 31 collieries proposed for closure in BC's statement of 13 October 1992 (which had precipitated the 'Coal Crisis'), had been 'saved' either by BC, or by the private sector through the lease/licensing process. However, as the table shows, this was not the whole story. In order to achieve a balance between supply and falling demand, BC had to cease production at eight collieries previously expected to continue, by a combination of closure, merger and 'care-and-maintenance'. In all essentials, BC had arrived at the same position as envisaged in October 1992, but a year later. Yet in spite of the large number of closures that took place in the year following the 'Coal Settlement' of March 1993, there was no public outcry, and little political difficulty. The Labour Party remained implacably opposed to the government's policy on the coal industry (as it had been since the Thatcher years) and initiated a further debate in the House of Commons in 27 October 1993 to protest against the new closures. But the government had achieved sufficient reassurance for its own party in the 'Coal Settlement', and won by 34 votes. The opposition of the Labour Party counted for nothing, so far as public policy on coal was concerned.

There were three factors in particular which facilitated the resumption of large-scale closures during the financial year 1993/4. First, after the statements by government and BC on 25 March 1993, setting out the 'Coal Settlement', there were no further big announcements which could act as a focus for political or widespread public protest (as had happened in October 1992). Second, in September/October 1993 BC reached an accommodation with the government whereby, in return for an undertaking by BC to continue to operate the pre-closure consultation process embodied in the MCRP (which BC had wished to replace with a new 'streamlined' procedure), BC was enabled to offer a supplementary redundancy payment of £7,000 per man (funded by government), on top of the payments announced in March

Table 6.1: Number of BC Collieries. March 1993 and May 1994

| The Coal Settlement | | Numbers at May 1994* | | | |
Category at 25.3.93	Numbers at 25.3.93	Closed	Leased/licensed	BC C+M(3)	BC Production
1. Ceased production Oct'92	10	7	3	-	-
2. Care and Maintenance	6	4	2	-	-
3. Continuing Production during 'market testing'	12	5	3	-	4
4. Early closure	2	2	-	-	-
5. Long-term development	1	-	-	-	1
Total planned to cease production 13.10.92	31	18	8 (2)	-	5
Total planned to continue 13.10.92	19	4 (1)	-	4	11
Asfordby New Mine	1	-	-	-	1
Total	51	22	8	4	17

Notes   (1) Includes 3 collieries in process of merger with long-life pits.
       (2) Includes 3 collieries where lease/licence not completed until later in 1994.
       (3) Collieries placed on 'Care-and-maintenance'. However the table excludes several units which had been under 'care-and-maintenance' for some years.

* Status of individual collieries taken from the Government Energy Report Vol. 1 'Markets in Transition' 1999

1993, on condition that closures were agreed at local level (i.e. avoiding national appeal). The Minister subsequently confirmed that these enhanced terms could be continued to April 1994 'or as long as consultations proceed'. In effect, this enabled BC once again to decentralise the whole closure process in a way which minimised both the political impact and the possibility of significant resistance by the unions. Thirdly, the lease/licence arrangement (whereby the site of a surplus BC colliery was leased to a private coal operator, together with a licence to work the coal) appeared to be sufficiently successful for the government to gain credit with its own party: of the 31 collieries originally proposed in October 1992 for early closure, 26 subsequently proved to be surplus to BC requirements, and of these eight were leased/licensed to the private sector. Even the eighteen collieries which finally closed had at least been given the opportunity to continue if a private buyer could have been found.

The outcome of the process was the granting of lease/licence to nine collieries during 1994, namely:

- Clipstone, Rossington and Calverton to RJB Mining.
- Markham Main, Trentham, Coventry* and Silverdale to Coal Investments (led by Malcolm Edwards)
- Hatfield and Betws to management buyouts.
  (*Coventry had not been part of the original closure programme, having been on 'care and maintenance' for some time)

In all, the lease/licence procedure was considered by the Conservative Party as a reasonable interim measure prior to full privatisation. However, it was not without concern to BC, which had feared that lease/licensing on a large scale might create competitive difficulties for the continuing BC pits in the run-up to BC privatisation. In fact, these fears proved to be largely groundless, for three reasons. First, the 'core contracts' between BC and the major generators, which were agreed in March 1993 (and finally signed in June 1993), and which secured the greater part of BC's output at predictable prices, were not available to lease/licence pits (which were treated as assets rather than businesses) which would have to accept much less favourable terms for sales to power stations. Second, although the potential capacity of these nine collieries was some 9m. tonnes per annum, their actual collective output in 1993 was little more than 2m. tonnes. Some had ceased production, and all would require time and money to bring output up to full capacity. Third, several of these collieries had been producing coal qualities designed primarily for the domestic and industrial markets.

Overall, therefore, the lease/licence process posed little threat to BC's main markets, or to BC's ultimately successful efforts to balance supply and demand in the run-up to privatisation.

The changes in the overall supply and demand for coal between 1992 and 1994 proved to be similar to that projected by BC, at the time of the Coal Review (See Table 6.2). As we have noted, the closures of deep mines and the resultant reduction in deep-mined output were spread over two years rather than one, made possible largely by the later phasing of the generators' stock lift as between 1993 and 1994. Nevertheless, between 1992 and 1994, the market for UK coal had been reduced by nearly 30m. tonnes, due mainly to lower power station coal consumption (arising from increased use of gas and improved performance of nuclear stations) and the large

Table 6.2: Supply and Demand for Coal 1992 to 1994. Million Tonnes

	1992	*1993*	*1994*
UK Coal Output [1]			
Deep mined	65.8	50.5	31.9
Opencast and other	18.7	17.7	17.9
Total	84.5	68.2	49.8
Stock increase/(decrease)	2.7	2.3	(4.7)
Sales of UK Coal [2]	81.8	65.9	54.5
UK Coal Consumption			
Power stations [3]	78.5	66.1	62.4
Other Markets	22.1	20.7	19.4
Total	100.6	86.8	81.8
Supplied by:-			
UK Coal [4]	80.8	64.8	53.3
Distributed coal stocks [5]	(1.1)	3.6	13.9
Imports etc [6]	20.9	18.4	14.6
Total	100.6	86.8	81.8

Notes (1) BC and other UK Producers
       (2) UK output +/- stock change
       (3) All power producers
       (4) Sales of UK coal less exports
       (5) Mainly at power stations
       (6) Including Miscellaneous and statistical difference

Source: *Digest of UK Energy Statistics 1999*, Tables 2.10 and 2.11 (DTI)

de-stocking by the generators. As a result, within two years, deep-mined output was more than halved. Excluding years of national strikes, the fall in output was far greater than in any other two-year period in the industry's Post-War history. But one of the necessary pre-conditions for successful privatisation had been met: output had been brought down to around the level likely to be required to meet demand in the period immediately following the sale of BC's mining assets.

The very rapid reduction in BC output in the run-up to privatisation was associated with an even more rapid reduction in BC colliery manpower, as a result, not only of colliery closures, but also of increases in labour productivity, which more than doubled between 1992/3 and the nine months to December 1994. (Output per manshift increased from 6.34 tonnes to 13.15 tonnes). Large reductions also occurred among other BC employees in non-mining activities, and in managerial and administrative staff (See Table 6.3).

Table 6.3: Changes in Numbers of BC Employees. End March 1992 to End December 1994. Thousands

	March '92	March '93	March '94	Dec. '94
Colliery industrial manpower	43.8	31.7	10.4	7.4
Other BC employees	14.3	12.5	8.5	5.7
Total	58.1	44.2	18.9	13.1

Source: *BC Annual Reports*

It is noteworthy that the overall reduction in the numbers of BC employees between March 1993 (at the time of the 'Coal Settlement') and privatisation in December 1994, was 31,000; almost precisely the number envisaged in BC's original closure announcement of 13 October 1992. As with deep-mined output, the effect of the 'Coal Settlement' of March 1993 on the numbers employed by BC was not to change the final outcome, but merely to delay it by a year or so. Furthermore, by December 1994 the colliery labour force had been reduced to the point where the potential problems (and costs) of redundancy in the period following the sale to the private sector were likely to be minimal. Another important pre-condition of successful privatisation had been satisfied.

The large increases in labour productivity together with the closure of higher cost pits, were also instrumental in reducing deep-mined operating costs. The increasing use of 'heavy duty' coalface machinery, retreat mining, roof-bolting and general management action to reduce manpower at continuing collieries as output was obtained from fewer high-output faces, all contributed to the sharp increase in efficiency. This process was also encouraged by the government's removal of the limitations to flexible working in the 1908 Act, and by greater use of incentive, bonus and overtime payments, so that by 1994, the 'basic wage' made up less than half of mineworkers' average weekly earnings. Between 1991/2 (the year before the Coal Review) and the nine months to December 1994 (when BC was privatised) average colliery operating costs fell by 27 per cent in 'real' terms (using the GDP deflator).

These improvements in costs enabled operating profits to be made during the transitional period between the 'Coal Settlement' of March 1993 and privatisation in December 1994, (although at a much lower level than in 1992/3) in spite of the much lower prices in the new contracts with the major generators operative from April 1993.

The overall financial results were affected by restructuring costs and other exceptional items (See Table 6.4). But in terms of the preparation for privatisation what mattered was not so much BC's overall financial results, but whether operating costs (particularly for deep mines) had been sufficiently reduced to enable potential buyers of BC to have the expectation of reasonable profits, given the predetermined prices in the contracts with the generators.

More specifically, deep mines had to be seen to be competitive against the prices envisaged for 1997/8, the last year covered by the contracts with the major generators. Average revenue for all sales was unlikely to diverge greatly from the prices expected for sales to power stations. Table 6.5 shows the average revenues by main market in the nine months to December 1994.

The five-year contracts with the generators provided for prices in 1997/8 of £1.33/GJ in September 1992 money, equivalent to about £1.40/GJ in 1994/5 money. During the nine months to December 1994, the fourteen BC collieries still fully operational in December 1994 (that is excluding Maltby and Asfordby, which were under development, and a number of collieries under 'care and maintenance') had average 'cash operating costs' (that is, excluding depreciation, but including routine expenditure on plant and machinery etc.) of £1.20/GJ, and only five collieries had costs in excess of £1.40/GJ (BC Management accounts). Thus, by 1994, average colliery costs had already been reduced sufficiently to yield prospective average operating profits of about 15 per cent of turnover in 1997/8, even before taking credit for expected further reductions in costs.

Thus, by the time of privatisation, BC's management of the industry

Table 6.4: Summary of BC Finance Results 1992/3 to 1994/5. £Million

	1992/3	1993/4	1994/5 (9 months to Dec'94)
Operating profit			
Deep mines	341	66	62
Opencast	160	80	41
Others	19	40	15
Total	520	186	118
Exceptional items [1]	(996)	(305)	(213)
Net interest charges	(112)	(119)	(90)
Extraordinary items [2]	-	-	662
Total	(588)	(238)	477

Source: *British Coal Annual Reports*

Notes:

(1) Make-up of exceptional items:

Social costs	(767)	(481)	(270)
Less grants	646	423	195
Net social costs	(121)	(58)	(75)
Write-down of fixed assets and stocks *	(636)	(84)	-
Other closure costs etc	(239)	(163)	(138)
	(996)	(305)	(213)

* Arising from new Contracts with generators

(2) Extraordinary items in 1994/5 : Net effect of:

Net assets transferred to successor coal companies on privatisation	(1262)
Liabilities transferred to Coal Authority	291
	(971)
Plus extinguishment of loans under Coal Industry Acts	1633
Total	662

Table 6.5: BC Average Revenues by Main Market. 9 Months to December 1994

	£/tonne	£/GJ
Power Stations	33.12	1.40
Industry	30.57	1.17
Domestic	55.18	1.90
Exports	24.32	0.90
Average	33.84	1.39

Source: BC management accounts

had delivered the three inter-locking preconditions of a balance between supply and demand, virtual completion of the programmes of colliery closures and redundancies, and operating cost levels capable of delivering acceptable profit levels after privatisation – all within the framework of the contracts with electricity generators already brokered by government. Prospective buyers of BC's mining assets were therefore largely freed from economic risks during the remaining duration of the contracts with the generators, rather more than three years after the planned date of sale, in that the vast majority of sales were secure, and at prices significantly higher than the current operating costs of the remaining deep mines (and opencast sites).

The Legislative and Administrative Process

But there were other potential risks to be taken into account, namely those of liabilities arising from past operations of British Coal, and from the structure of ownership which would arise after the transfer of ownership to the private sector. These were matters in the hands of the government, rather than British Coal, to be dealt with in the legislative and administrative processes of privatisation itself.

The legislation required for coal privatisation was embodied in the Coal Industry Act 1994, and given Royal Assent on 5 July 1994. As Lord Strathclyde, the government minister responsible for steering the Bill through the House of Lords, said, 'The main purpose of the Bill is privatisation. The essence of privatisation is to pass the assets and liabilities from the public sector to the private sector. That is what privatisation is all about' (Hansard H. of L. 21/6/94 column 216).

The Act contained the provisions necessary to enable BC's mining operations and other assets to be sold, and for the Corporation to be dissolved as a legal entity in due course.

The Act also provided for the establishment of a new public-sector body, the Coal Authority, with functions which would include:

(i)   owning the nation's reserves of unworked coal (which were previously owned by BC) and granting access to those reserves by licensing coal mining (an activity previously carried out by BC in respect of small deep mines and opencast sites);

(ii) responsibility for the physical legacy of previous mining where this was not transferred to the private sector;

(iii) making available mining records and geological information, including those transferred from BC.

Although the Coal Authority would take over the ownership of UK coal reserves from BC, it would be independent of the coal-mining industry and would not, itself, be allowed to carry out mining for commercial purposes. Under the new licensing regime (operated by the Coal Authority from November 1994) it was unlawful to mine coal without a licence, although those private-sector mining operators previously licensed by BC had the option of maintaining their previous licences for their duration, or applying for fresh licences. During the process of privatising BC, the Act gave powers to the Secretary of State to act on the Coal Authority's behalf in issuing licences for the sale of BC to preferred bidders (see below); but subsequent licences would be available on an open-access basis, subject to the Authority satisfying itself as to the applicants' financial resources and expertise.

The Act made it clear that grants payable to BC under previous legislation would not be transferable to successor companies; and the Secretary of State was empowered to extinguish outstanding loans made to BC.

Other provisions of the Act included:

(i)   Confirmation that the Health and Safety Executive, through Her Majesty's Inspectorate of Mines, remained the safety enforcement authority.

(ii) Government guarantees of the solvency of existing pension schemes and the payment of pensions updated annually in line with the retail price index.

(iii) Government safeguards to existing entitlements to concessionary fuel to former BC employees and their dependants.

(iv) Clarification that methane in coal seams was vested in the Crown, and subject to the oil and gas licensing and fiscal regime. The Coal Authority was required to establish a regime for exploration and development of coal-bed methane, but without jeopardising existing or future coal mining.

The Act also had important implications on the issue of liabilities. Given that the sale of BC was as a going concern, this involved purchasers taking over responsibility for contracts and liabilities outstanding at the time of transfer. In the case of contracts for the sale of coal, notably contracts with electricity generators, this was no burden: precisely the opposite. Indeed, as we have already observed, such value as the successor companies had depended largely on the unexpired part of these contracts. On the other hand, the way liabilities were dealt with would have a crucial effect on the confidence of potential purchasers, and their assessment of future risks. There were a number of areas of concern.

Firstly, claims for damage arising from subsidence following underground mining operations could be both large and unpredictable. The Act provided that purchasers would be held responsible for subsidence only in the 'area of responsibility' set out in their licences. During the passage of the Bill, the government clarified this to mean only those areas likely to be affected by mining after March 1994, and, further, indemnified successor companies against claims *within* these areas, where BC had already admitted liability. Claims against subsidence damage arising from underground workings carried out before March 1994 would be the responsibility of the Coal Authority. Other liabilities arising from past workings such as problems arising from abandoned shafts, also became the responsibility of the Coal Authority.

Secondly, given that BC was privatised as a 'going concern', the Transfer of Undertaking (Protection of Employment) (TUPE) Regulations applied. Terms and conditions of employment enjoyed by employees at the time of sale were thus to be carried over by successor companies. These included recognition of trade unions, and previous agreements with trade unions, while not precluding subsequent renegotiation of wages. In order to prepare for this, BC in early 1994 introduced a package, whereby mineworkers would receive a lump sum of £6,000 if they would accept a revised employment contract which would support flexible working arrangements, and which sought to 'contractualise' the redundancy payments scheme which hitherto had been 'ex gratia'. The proposals were for three weeks' pay, up to a maximum of £300 per week, for each year of service, up to a maximum of thirty years; thereby providing for redundancy payments of up to £27,000 per man. As it proved impossible to secure the agreement of the main unions to this package, it was put to individual members for decision by August 1994, by which time 79 per cent of mineworkers and over 90 per cent of other relevant grades had accepted the package. These contractualised redundancy arrangements were to apply for four years to March 1998, and were specifically linked to the duration of the coal sales contracts with the major generators. As a result, a very generous redundancy scheme, hitherto largely funded by government grants to BC, would fall to the successor companies. In practice, this potential problem for purchasers was greatly mitigated (as we have seen) by the fact that nearly all of the workforce reduction required for viable operation in the period up to March 1998 had already been carried out by BC at government expense. Further, TUPE did not apply to those former BC employees who had refused to accept BC's package, to 'outside' contractors employed on tunnelling and other development work, or to mineworkers employed at collieries already transferred to the private sector under the lease/licence arrangements. The provisions of TUPE would also effectively not apply to those limited number of collieries under 'care and maintenance' to be offered for sale (since they had very few current employees, other than those involved in the maintenance operations).

Thirdly, during the passage of the Bill, the government clarified the position whereby responsibility for all health and injury claims with respect to past service with BC would be retained in the public sector and accepted by the government. This included responsibility for claims with respect to past service by BC employees who transferred to successor companies. Responsibility for claims arising from *future* service would rest with the successor companies. But as the claims would almost all refer to past service (even for transferred employees) this clarification greatly reduced the anxiety of potential purchasers on this score.

Overall, the liabilities issue had been resolved in a way which provided as clean a break with the past as was practicable, given the complexity of the issues involved, and for nearly all of these liabilities to remain with the public sector. Another of the essential preconditions for successful coal privatisation had been met.

The New Structure of the Privatised Coal Industry

Finally, there was the issue of structure. We have already noted above that by the time of the 'Coal Settlement' of March 1993, the government had determined BC would not be privatised as a unitary entity. Indeed, this had been clear as far back as mid-1991 during discussions between BC and N M Rothschild, the government's main advisor. The Board of BC were concerned to avoid what they saw as

the dangers of excessive fragmentation of the industry after privatisation, on the grounds that a fragmented industry would be even more
exposed to the market power of the duopoly of National Power and
PowerGen than it was already. Neil Clarke warned the Minister, Tim
Eggar, of these dangers in July and again in September 1993, but by
then the government had decided in principle to accept the
recommendations of Rothschilds for a five company structure. Some
BC non-Executive Directors suggested that BC might need publicly to
point out the risks involved in this degree of fragmentation, but in the
event BC did not make any dissenting comment when the Minister
announced, on 21 September 1993, that the government intended to
offer for sale five regional businesses, based on Scotland, Wales, North-
East England and the Central Coalfield (which would be offered in
two separate parts). Potential purchasers would be given the opportunity to bid for one or more of the regional businesses, up to and
including all five packages, with the bids to be assessed on their merits.

This approach was notable for a number of reasons. Firstly, the
government had clearly accepted that, given the extent to which
individual colliery performances fluctuated, a degree of aggregation of
assets (rather than sale on a colliery-by-colliery basis), was required to
create the conditions for sustainability after privatisation. Secondly, it
indicated that the government wished the broad structure of the
industry after privatisation to be determined by a competitive tendering
process which could be characterised as a 'market solution'. Thirdly,
and most interestingly, the option of multiple bids, including at least
the possibility that all five regional businesses could fall into the same
ownership, implied that in the last analysis the government placed
more importance upon the successful completion of privatisation, in a
way which maximised proceeds to the Exchequer, and on the abolition
of British Coal, than on the creation of intra-industry competition.

The proposed sale structure was set out in more detail in the
'Preliminary Memorandum': Sale of the Coal Mining Activities of
British Coal Corporation, issued by Rothschilds on 13 April 1994,
which gave general information, primarily of the five Regional Coal
Companies which would be offered for sale after the Coal Industry Bill
had received Royal Assent. These Regional Coal Companies were not
actual operating entities, but portfolios of assets consisting of (named)
operating collieries and opencast sites (together with associated 'disposal
points') and prospective opencast sites, BC's coal stocks at the date of
sale, and the relevant share of BC's contracts with electricity generators
and other customers (Table 6.6).

In addition, but not as part of the five Regional Coal Company

Table 6.6:   Assets of Five Regional Coal Companies Offered for Sale, 1994.
Million Tonnes

| | | Deep Mines | | | Opencast | |
	No. of Pits	1993/4 output	Reserves	1993/4 output	Reserves Existing	Prospective
Central North	9	16.3	213	2.0	5	55
Central South	6	9.0	220	2.4	8	40
North East	-	-	-	2.2	12	30
Scotland	1	1.5	21	3.7	17	35
South Wales	-	-	-	2.4	8	45
	16	26.8	454	12.7	50	205

Notes: Central North deep mines include 5 collieries in Selby complex, and Maltby
(then 'under development')
Central South deep mines include Asfordby New Mine (under development)
Deep-mine reserves are 'Classified'
Opencast reserves : 'Existing' at currently contracted sites
'Prospective' at potential sites identified by BC

packages, five collieries currently on 'Care and Maintenance', and two
further collieries due to cease production during 1994, were offered
for sale. (This process excluded nine collieries already transferred to
private operators by BC under the 'lease/licence' arrangements, or
under negotiation for such transfer).

The sale of opencast sites (with associated disposal points) was
straightforward only in the case of existing sites already contracted,
and with planning permission. The 'prospective' opencast sites were at
various stages in the processes of obtaining rights over land and
planning and other consents. Thus, although purchasers would receive
conditional operating licences for these 'prospective' sites, the grant of
planning consents would remain a matter for the planning authorities,
with no presumption that such consents would be forthcoming.

The Preliminary Memorandum also set out the planned timetable
for sale, whereby applicants who 'pre-qualified' would be given access
to detailed BC records on individual collieries and other assets, with
subsequent bids being made by September, and the sale to be
completed by end December 1994.

Following the issue of the Preliminary Memorandum, 25 potential
bidders 'pre-qualified', of whom eighteen put forward bids for one or
more of the Regional Coal Companies. The pattern of bids was of
considerable significance.

No major multi-national company with energy interests made a bid.

Only one, RTZ, pre-qualified, but did not proceed to the bidding stage, in spite of active government encouragement. Efforts to interest US mining companies failed. Notwithstanding the progress that had been made by British Coal to increase productivity and reduce costs, and the government's efforts to limit future liabilities for prospective purchasers, international and overseas companies clearly did not regard the UK coal industry as an internationally competitive business. The two major generators, National Power and PowerGen both 'pre-qualified' for all five Regional Coal Companies, but did not bid. They were able, however, as a result of this process to gain access to the detailed information on the collieries, opencast sites and coal reserves. This was probably a necessary concession to obtain the generators' co-operation in splitting the coal supply contracts for British Coal between the five Regional Coal Companies.

Nearly all the bidders fell into two groups. First, opencast contractors who had derived most of their mining income from operating sites under contract for BC (although some also had small deep or opencast mines licensed from BC). Under the previous regime, whereby BC was the owner of all but the smallest opencast sites, and the operators were selected by BC under competitive tender, a significant part of the economic rent of production (particularly when sold to the generators under the high prices in BC's supply contracts), went as profit to BC rather than the opencast contractors (although BC had had the serious task of securing planning permission to work sites). No doubt, a number of contractors saw the opportunity of an enhanced share of the economic rent after privatisation. Most bids from opencast contractors were for the Regional Coal Companies of Scotland, North East and/or South Wales, which were coalfields with little deep-mining activity. There was, however, one important exception: RJB Mining bid for all five Regional Coal Companies.

The second group of bidders were led by current and former employees of BC, usually with a strong interest in the continuation of deep-mining activities. The prospective management buy-outs of English Coal and Northern Coal (led respectively by Bob Siddall and Alan Houghton, senior BC mining engineers), together with Coal Investments (led by Malcolm Edwards, the former BC Commercial Director) in collaboration with the UDM, made the majority of bids for the two central companies, which had virtually all the deep mines for sale. Current and former BC employees also figured prominently in Celtic Energy which bid (successfully) for the South Wales package; and Coal Investments was the largest single element in the consortium which made up Mining (Scotland) which bid for the Scottish assets.

To all intents and purposes therefore, the whole bidding process was confined to partners already participating in the UK coal mining industry, either as opencast contractors or employees of BC. But there was one very notable absence: the main body of the existing senior management of BC. In other privatisations of nationalised energy industries (British Gas, the major electricity generators, Regional Electricity Companies and National Grid Company), the incumbent senior managements had generally been major beneficiaries of privatisation; but in the case of BC, the management which had successfully completed one of the most radical restructuring plans of any major industry, was to be discarded. The cultural revolution would be complete.

On 12 October 1994, only a month after the closing date for bids, the government announced the 'preferred bidders' for the Regional Coal Companies and certain care-and-maintenance collieries (*). These were:-

- RJB Mining for Central North, Central South and North East Regional Coal Companies, together with Thorne* and Ellington*;
- Celtic Energy for South Wales;
- Mining (Scotland) for Scotland (which was to trade as 'Scottish Coal')
- Coal Investments for Annesley/Bentinck*;
- Tower employee buy-out for Tower*.

No complying bids were received for the three other care-and-maintenance collieries, which duly closed.

While separate sale of the Scottish and South Wales coalfields had generally been expected, the proposed sale of virtually all BC's mining activities in England (including nearly all the deep-mines) to RJB Mining caused widespread surprise among commentators on several grounds. First, the government had earlier been very unsympathetic to the view put forward by BC that the greater part of its mining operations should be sold as an entity, particularly in England, in order to mitigate the serious imbalance of market power in favour of National Power and PowerGen; and the government and its advisers (helped by BC) had spent much time and effort finding ways of dividing BC's mining activities into the five Regional Coal Companies.

Second, the decision largely excluded Coal Investments which, with the UDM, had bid for Central North and Central South. This partnership was widely expected to secure at least Central South, where the UDM predominated – not least because of the perceived 'debt of gratitude' which the government had previously expressed to the UDM for standing firm against the NUM during the 1984/5

NUM strike. But with the passing of Mrs Thatcher from the scene, and the effluxion of time, this was no longer a decisive factor. Third, there was some scepticism as to the viability of the RJB bid. RJB initially bid £914m. for the three England Regional Coal Companies, which it seemed that other bidders valued at £600m. or less. Although, with subsequent agreement with government, the RJB bid was scaled down to £815m. (reflecting revised coal stock valuations and other adjustments to the original offer), there was much comment that the company appeared to be taking too optimistic a view of market prospects. Moreover, a prospective annual tonnage of well over 30m. tonnes was vastly greater than the company's output in 1993 of little more than 2m. tonnes, nearly all of which was opencast under contract to BC; whereas most of the newly acquired tonnage was in the form of large deep mines (particularly the five mines making up the Selby complex) of which RJB had little previous experience until the acquisition and re-opening of three surplus mines in 1994 under the lease/licence arrangements from BC.

Nevertheless, the government claimed to be pleased with the outcome. Tim Eggar's statement on 12 October 1994 (DTI Press Release P/94/603) stated that 'we received a range of very good bids for the regional coal companies ... There will now be a period of detailed negotiation with the preferred bidders, with a view to completing the sales by the end of the year.' The statement did not give any indication of the reasoning behind the selection of preferred bidders: but it did not need to do so. It was inherent in the competitive bidding process that the sums offered would be the major determinant of the government's choice, always provided that they were satisfied (on the basis of the advice particularly of Rothschilds) that the selected bidders would be able, both financially and technically, to sustain operations after purchase. And in the case of RJB Mining, the bid was so far ahead of other bids in terms of the sum offered for the English coalfields that a more complex balancing of other criteria would not have been appropriate. Within the framework set by the 'Preliminary Memorandum' in April 1994, the government had been given a 'market solution', which would raise some £959m.: £815m. from RJB Mining, £95m. from Celtic Energy, and £49m. for Mining (Scotland): a total sum close to the book value of assets, less liabilities, passed to successor companies, as shown in BC's Report and Accounts for 1994/5 (£971m.: Note 19 to accounts) but significantly, more than the government had been led to expect by its advisors.

Between 12 October and 31 December 1994, the completion of the privatisation of BC's mining activities was in large measure in the hands of Richard Budge, the Chief Executive of RJB Mining, who showed great energy and resourcefulness, both in raising the necessary finance, and in setting up the organisation required to effect a transition from a relatively small, mainly opencast, business into one of the world's largest privately owned coal-mining companies.

RJB had been created in 1992 by Richard Budge, who led a management buy-out of the mining activities of the AF Budge group. RJB was floated on the London Stock Exchange in May 1993, and in December 1993 acquired the private mining business of Youngs. But in 1993, the company was still a small-scale producer of 1.8m. tonnes of opencast coal (of which 1.6m. tonnes was under contract from BC) and 0.3m. tonnes from the licensed Blenkinsopp deep mine. In 1994, the Company acquired the Monckton smokeless fuel plant, and three BC mines under lease/licence (Clipstone, Rossington and Calverton), which had been re-opened by the end of 1994, and with a combined potential output of some 2m. tonnes. But, even when these newly acquired deep mines were fully operational, the total coal production of the Company would still be only a small fraction of the output envisaged after the sale of BC's English mining assets (which RJB termed 'English Coal'). Yet, despite the massive scaling-up of the Company required, the necessary financial resources were secured. On 9 December, a syndicate of Banks had agreed, subject to certain conditions, to enter into a £494m. loan agreement with RJB. In November, the Company issued what was effectively its prospectus for 'English Coal' to raise the balance of £385m. of its new capital requirements from the issue of 125m. new ordinary shares at 320p per share. The offer was successful, and trading began on 30 December 1994, when the sale of 'English Coal' to RJB was completed. This had been a great achievement for Richard Budge.

At the time of sale of BC's Scottish assets to Mining (Scotland) – soon to be known as Scottish Coal – the new owners were a syndicate of institutional investors (Murray Johnstone, Waverley Mining, Northern National Resources) 63.7 per cent, Coal Investments 22.9 per cent and employees 13.4 per cent.

Celtic Energy, a management buyout led by Bryan Riddlestone, previously BC's Opencast Manager in South Wales, acquired BC's South Wales mining activities.

In addition, Coal Investments had secured six former BC collieries (five under lease/licence); and management/employee buyouts had secured Hatfield, Betws and Tower collieries. By early 1995, therefore, the structure of the privatised UK coal industry was clear as well as the likely pattern of production (Table 6.7).

Table 6.7:   Structure of Privatised Coal Industry by Company. Early 1995. Likely Pattern of Production. Million Tonnes

	Deep Mines	Opencast	Total
RJB Mining	28	7	35
Mining (Scotland)	2	3	5
Celtic Energy	-	2	2
Coal Investments	2	-	2
Other	2	4	6
Total	34	16	50

This summary illustrates the dominance of RJB Mining, with over 80 per cent of the deep-mined output, and 70 per cent of total UK coal output – and with 29m. tonnes out of the 30m. tonnes per annum of the coal supply contracts with National Power and PowerGen.

The 'Ultimate Privatisation' Achieved

The 'ultimate privatisation' had been accomplished, even if the industry's structure fell some way short of a fully competitive ideal. The completion of the sale of the Regional Coal Companies, and the care-and-maintenance collieries and the earlier lease/licensing process, represented a major legislative and administrative achievement, made possible by the government's political will to see the process brought to a successful conclusion, the expertise of the government's advisors (particularly Rothschilds, and Clifford Chance) and the energy of the new owners (particularly Richard Budge). But British Coal also deserved much of the credit, as the Minister, Tim Eggar, acknowledged (statement 30 December 1994; DTI Press Release P/94/800). The untangling of the industry's statutory and historic liabilities, the separation of mining from non-mining activities, the allocation of assets and liabilities into the various packages offered for sale, the revision of terms and conditions of employment, the determination of property rights: all this was carried out for the most part by those made redundant (albeit under generous terms) on completion of the task. At the same time, BC had delivered (as we have seen) accelerating levels of productivity and cost reduction and a balance between supply and demand, and what appeared at the time to be a potentially competitive business. As Neil Clarke, the BC Chairman, said in effectively the last BC Annual Report: 'When the debate about the structure and ownership of the industry has become only a distant memory, the creation of a competitive business will have offered coal the best prospect of enduring success and provided the most fitting testimony to the endeavour of half a century' (*BC Annual Report 1994/5*, p. 8).

The government also had other reasons for satisfaction. The sale of BC's mining assets represented the completion of the 'political agenda' on coal: the 'ultimate privatisation' had been achieved.

The public rationale for coal privatisation had been set out in the White Paper of March 1993. A similar line was taken by the Energy Minister, Tim Eggar, in the DTI statement of 30 December 1993: 'I am convinced that the privatisation represents the best prospect for achieving the largest economically viable coal-mining industry in the longer term while ensuring that the taxpayer gets value for money.'

It is worth noting what the government's public rationale did *not* say. There was no explicit claim that coal privatisation was essential to confer benefit to electricity consumers; although the sale was conducted in such a way as to obtain the best available price for the taxpayer, there was no suggestion that a major reason for the sale was to raise money for the Exchequer; and, finally, the use of trade sales meant that the government's objectives did not include wider share ownership by the general public, as had featured in other privatisations. Rather, the government's public position was that the main beneficiary of privatisation would be the coal industry itself.

But this reasoning was far from clear. While it may be argued that the momentum for the large increases in productivity and closure of surplus, uneconomic capacity depended to a considerable degree on the preparation for privatisation within a tight timetable, this process could have occurred in a politically acceptable way only with government financing of generous redundancy in the public sector. These were the preconditions, not the consequences of privatisation. Also the amendments to working time and the introduction of flexible working practices were achieved while BC was still in the public sector. Any 'constraints of public ownership' might be expected to centre on the inhibition of investment arising from controls on the public-sector borrowing requirement. But this was no longer a real issue: major investment was unlikely to feature in the plans of the successor companies. The fundamental difficulties still facing the UK coal industry arose primarily, not from the 'constraints of public ownership', but from the rapid rise in the use of natural gas in power generation; the risks and uncertainties inherent in the coal market, particularly the availability of cheap imports; growing environmental concerns; and the difficult geology of the UK coalfields.

None of this amounts to a case *against* the privatisation of British Coal. Indeed, as the oil and gas production industries, and electricity and gas distribution networks were already in private ownership, there was no obvious reason to suggest that the coal industry was so different that it needed to remain in public ownership. Rather, what we seek to argue here is that, even in 1994, it was very doubtful that the act of transfer from public to private ownership would signal, as the government and its supporters appeared to claim, a renaissance in the industry's fortunes.

As the political rhetoric, both for and against public ownership, was left behind, the privatised coal industry was still faced with uncomfortable and unresolved long-term problems.

CHAPTER 7
THE PRIVATISED INDUSTRY: UNRESOLVED PROBLEMS

Early Success and Muted Politics

Developments immediately following the completion of the sale of BC's mining assets at the end of December 1994, seemed to augur well for the newly-privatised UK coal industry. Compared with 1994, 1995 saw an increase in deep-mined output and total sales of UK coal; and although UK coal consumption continued to fall, the collective market share of UK producers increased significantly as the huge coal stock-lift by the generators began to level out. (See Table 7.1)

The results of RJB Mining, which represented some 70 per cent of the privatised coal industry, were particularly encouraging. The Company's sales to power stations were over 6m. tonnes more than laid down in the inherited 'British Coal' contracts. This additional coal requirement by the generators arose mainly from delays in the build-up of output at a number of new CCGT stations at a time of greater than expected electricity demand; so that, although coal consumption at power stations fell in 1995, the fall was less than previously anticipated. RJB deep-mined output was higher than BC's output from the same collieries in 1993, including the effect of rebuilding the output of collieries acquired from BC under the lease/licence arrangements before full privatisation. (See Table 7.2)

The financial results were outstanding. Not only did the company achieve a remarkably smooth transition from 1994 (before most of the BC assets were acquired) with a twelve-fold increase in turnover and a ten-fold increase in profit; but the operating profit margins on turnover (13.8 per cent on deep mines and 20.4 per cent on opencast) were very healthy for a mining business. Moreover, the profit before tax of £173m. compared with the prospectus projection of £119m. for 'English Coal' (that is those BC mining assets acquired on privatisation in December 1994). The strong operating cash flow (which greatly benefited from the 4.8m. tonnes of stock reclaimed for sale as a result of the higher than expected requirements of the generators) enabled the accelerated repayment of £313m. of bank debt. The financial performance also allowed the granting of shares to employees through the RJB Mining Sharesave Trust, set up in April 1996. (See Table 7.3)

Table 7.1:   UK Coal Output and Consumption. 1994 to 1999. Million Tonnes

	1994	1995	1996	1997	1998	1999
UK Coal Production						
Deep-mined	31.9	35.1	32.2	30.3	25.3	20.9
Opencast etc	17.9	17.9	18.0	18.2	15.9	16.2
Total	49.8	53.0	50.2	48.5	41.2	37.1
Stock lift/(increase)	4.7	4.2	2.9	(0.6)	0.2	(0.4)
UK coal sales [1]	54.5	57.2	53.1	47.9	41.4	36.7
Coal Consumption						
Power Stations	62.4	60.0	55.4	47.2	48.5	40.5
Other Markets	19.4	16.9	16.0	15.9	14.6	15.0
Total	81.8	76.9	71.4	63.1	63.1	55.5
Supplied by:-						
UK coal [2]	53.3	56.3	52.1	46.8	40.4	35.9
Distributed coal stock lift/(increase) [3]	13.9	4.4	0.9	(3.3)	1.0	(0.6)
Imports etc [4]	14.6	16.2	18.4	19.6	21.7	20.2
UK coal's share of UK coal market	65%	73%	73%	74%	64%	65%

Notes: (1)   Production plus/minus undistributed stock change
(2)   UK coal sales less exports
(3)   Mainly power station coal stocks
(4)   Includes statistical balancing item

Source: *Digest of UK Energy Statistics* (DUKES) 2000

Although in its Report and Accounts for 1995, the Company warned that it did not expect the additional sales of power station coal to be repeated in 1996, it gave an up-beat message for the future, drawing attention to the upward revision of its assessment of coal reserves by 138m. tonnes, and the £300m. which had been expended during the year on underground mining development, including 160km. of roadway drivage. The City was greatly impressed by the Company's performance, and the abilities of Richard Budge, the Chief Executive: RJB's share price increased from the offer price of 320p in December 1994 to a high of 625p in the first half of 1996.

Scottish Coal and Celtic Energy, which had acquired the Scottish and South Wales mining assets of BC on privatisation in December

Table 7.2:   RJB Mining Output and Sales 1995 to 1999. Million Tonnes

	1995	1996	1997	1998	1999
Output					
Deep mines	29.9	27.7	25.3	19.8	17.5
Opencast	7.2	7.2	6.5	5.8	5.0
Total	37.1	34.9	31.8	25.6	22.5
Stock lift/(increase) *	4.8	2.7	(0.6)	0.3	-
Total sales	41.9	37.6	31.2	25.9	22.5
of which: Power Stations	35.7	33.4	26.8	22.5	19.4
Other Markets	6.2	4.2	4.4	3.4	3.1

*   including balance between sales and production

Source: RJB Mining

Table 7.3:   RJB Mining: Summary of Financial Results 1995 to 1999. £Million

	1995	1996	1997	1998	1999
Turnover	1461.3	1308.5	1124.7	822.5	699.2
Operating profit:					
Deep mines	166.8	149.5	137.5	25.7	12.5
Opencast [1]	43.1	52.4	44.3	32.3	11.2
Manufactured Fuel etc.	1.4	1.3	1.1	0.7	0.1
Total	211.3	203.2	182.9	58.7	23.8
Net interest etc.	(38.2)	(14.0)	(11.8)	(18.6)	(153.8)[2]
Profit/(Loss) before Tax	173.1	189.2	171.1	40.1	(130.0)[2]
Net cash flow before financing	319.8	214.1	154.1	19.1	15.0
Operating profit margin on sales	14.5%	15.5%	16.3%	7.1%	3.4%

(1) includes opencast operations under contract to other coal owners
(2) including exceptional item of £131 million, reducing the carrying value of colliery assets

Source: RJB Mining

1994, also made good, if less spectacular progress in 1995, with aggregate profits of £27m. in the 15 months to March 1996 (see Table 7.5).

The management buy-outs at Betws and Hatfield collieries appear to have had a successful year, and Goitre Tower Anthracite Ltd (the miners' buy-out of Tower colliery) made £3.5m. profit, to much acclaim.

In marked contrast, particularly to RJB Mining, Coal Investments (CI), (led by Malcolm Edwards, BC's former Commercial Director) had a disastrous year, culminating in the suspension of share dealings and the company being taken into administration in February 1996. The difficulties faced by Coal Investments only served to highlight the differences in its circumstances compared to those of RJB. Unlike RJB, Coal Investments inherited no slice of the very favourable 'British Coal' contracts with the generators; so that, unlike RJB, it did not have most of its market guaranteed at premium prices, and had to seek sales at lower margins. Further, whereas nearly all the RJB deep-mine capacity was fully operational and had the benefit of BC's previous coal face development expenditure, four out of the five BC collieries acquired by CI had previously ceased production and were being held on a care-and-maintenance basis, requiring substantial capital expenditure for redevelopment of new coal faces before any financial return. Moreover, unlike RJB Mining, CI had no low-cost opencast production.

In spite of the failure of CI, 1995 had been a good year for the privatised industry. The government had been quick to draw attention to the increase in deep-mined output, and in its 1996 Energy Report (Vol.1 para 11.2) referred to the 'impressive achievements' within the industry. Yet despite the government's satisfaction, and the City's euphoric view of RJB, the situation within the industry caused little public comment and remained politically very low key.

But any signs of a new dawn for the privatised coal industry proved to be short-lived. Although in 1996 and 1997, the proportion of the UK coal market supplied by UK coal remained stable, UK coal consumption contracted sharply, and UK coal production and sales resumed their long-term decline (Table 7.1). In the crucial power station sector, there was a sharp increase in the use of gas at the expense of coal, as an increasing number of CCGTs initiated under the 'dash for gas' became fully operational, as had been foreseen at the time of the 'Coal Crisis' and Review of 1992/3. This was the main reason for the fall of 15m. tonnes in the annual power station coal consumption between 1994 and 1997 (Table 7.4).

Table 7.4:   Fuel Use by Major Power Producers. 1994 to 1999. Million Tonnes
Coal Equivalent

	1994	1995	1996	1997	1998	1999
Coal	60.7	57.9	53.4	45.3	46.6	39.0
Oil	6.0	5.1	5.0	2.3	1.3	1.3
Gas	15.3	18.9	25.1	31.6	32.8	38.9
Nuclear	35.8	35.0	36.6	37.7	38.0	36.5
Hydro/renewables *	0.8	0.8	0.6	0.7	0.9	0.9
Net imports (French link)*	2.5	2.3	2.4	2.4	1.7	2.0
Total	121.1	120.0	123.1	120.0	121.3	118.6
Coal's share	50%	48%	43%	38%	38%	33%

*   Based on electricity output, not coal displacement.

Source: DUKES 2000: Table 5.4 (derived).

The fall in both output and sales of UK coal between 1995 and
1997 was concentrated on RJB Mining's supplies to power stations
and the company's deep-mined production, which declined by some
9m. tonnes and 5m. tonnes respectively (See Table 7.2). Although RJB
inherited contracts with the major generators for an average of some
29m. tonnes per annum to March 1998, these were to some extent
rescheduled to provide for higher sales at the beginning of the period
(arising from higher than expected demand by the generators) but with
the consequence of a lower take in 1997.

In spite of the resumption of downward trends in coal sales and
output, RJB continued to show good financial results. Over the three
years 1995 to 1997, pre-tax profits averaged £178m., and operating
margins were maintained by cost reductions in real terms which
matched the reductions in revenue per tonne arising mainly from the
falling real prices written into the contracts with the major generators.
Net profit also benefited from the reduction in debt in 1995; and
although cash flow declined with the reducing stocklift, it still remained
healthy (Table 7.3). Over the same period, Scottish Coal and Celtic
Energy were more mixed in their financial fortunes; but collectively,
pre-tax profits over the three years 1995/6 to 1997/8 averaged some
£12m. (Table 7.5).

The overall picture, therefore, for the three years 1995 to 1997, is
one of a privatised industry with an aggregate pre-tax profit of some
£200m. a year, and, in the case of the largest producer (RJB Mining),
with the contraction in sales and output made tolerable by the under-

Table 7.5: Scottish Coal and Celtic Energy Summary of Financial Results. 1995/6 to 1998/9. £Million

	1995/6 (15 months)		1996/7		1997/8		1998/9	
	SC	CE	SC	CE	SC	CE	SC	CE
Operating profit/(loss)	8.6	18.7	(9.5)[1]	2.2	14.7	12.6[2]	(1.0)[3]	4.8
Net interest etc	(1.0)	0.6	(0.8)	0.4	(0.9)	(8.9)	(0.1)	(4.6)
Pre-tax profit(loss)	7.6	19.3	(10.3)	2.6	13.8	3.7	(1.1)	0.2

(1) The substantial loss at Scottish Coal in 1996/7 was due to serious but temporary loss of output at Longannet (the Company's sole deep mine)
(2) Includes receipt of payment of £5m. to Celtic Energy in recompense for anti-competitive practices by German anthracite producers, representing the resolution of a complaint to the European Commission.
(3) The operating loss at Scottish Coal in 1998/9 reflected lower prices and a £6.6m. loss at Longannet.

Source: Annual Reports: Mining (Scotland) Ltd and Celtic Energy Ltd.

pinning of 80 per cent of the business by the take-or-pay contracts with the major generators, inherited from British Coal.

During 1995 and 1996, there was a virtual absence of anything other than local public interest, or therefore any national political concern, about the coal industry's affairs. Colliery closures were few and far between, and the NUM, which had once dominated the politics of the industry, had become almost invisible. On 29 December 1994, at the point of the final sale of BC's mining assets, Arthur Scargill wrote to Kevan Hunt, BC's Director of Employee Relations, castigating BC for its part in the 'criminal butchery' of the coal industry, and stating that 'from the first day of privatisation, the NUM will intensify the campaign to renationalise the mining industry, and do everything in our power to restore the dreams and aspirations of our forebears who brought about a successful nationalised coal industry' (quoted in *Coal UK* (CUK) 37/18). Scargill seemed to think that this cause could best be advanced by staging another strike as soon as possible. By May 1995, he had obtained an 83 per cent vote in favour of selective 24-hour strikes in response to a proposal by RJB (later abandoned) for a freeze in basic wage rates. However, this vote represented only about 50 per cent of the NUM membership, and a much smaller proportion of the total labour force, as the UDM did not support the action (CUK 41/10). This threat was averted by RJB gaining an injunction for the union to reballot, and the action collapsed (CUK 42/9).

Scargill's next attempt at strike action was not until the end of 1996, when the NUM held a ballot for support of action for a 'substantial' pay rise, covering members working for RJB mining, and mining contractors. The result was a narrow majority in favour of action with 1,435 in favour and 1,231 against (CUK60/15). Once again RJB obtained an injunction requiring a reballot due to alleged irregularities in the first count, and a further ballot was launched (CUK 61/10), but to no effect. Almost as significant as the failure of those attempts to renew the former militancy of the workforce was the smallness of the numbers involved: a ballot with a total vote of less than 2,700 can be compared with a colliery labour force (over-whelmingly NUM members) of over 181,000 in March 1984, at the beginning of the Great Strike.

Early Signs of Problems to Come

Yet the return of politics to the affairs of the coal industry was not far away. The impending expiry of the contracts with generators, scheduled for March 1998, began to throw its shadow back into 1997. Minds began to turn to the prospects of the industry thereafter, and, after Labour's landslide victory in May 1997, to the implications for the industry of the change of government.

RJB Mining, which had some 85 per cent of the UK's deep-mine capacity, had particular reason for anxiety over the forthcoming end to the contracts, which continued to provide take-or-pay security for about 85 per cent of its total output, at premium prices. The main factor which reduced the total power station coal market by 15m. tonnes between 1994 and 1997 – the building of new CCGTs – looked set to continue to depress the coal market after 1998. Between the White Paper coal settlement in March 1993 and March 1996, over 10GW of new gas-fired stations were given Section 36 Consent by Government, and, although all might not proceed according to plan there was the prospect of a further loss of well over 20m. tonnes of coal business by the end of the 1990s.

The outlook on prices was also disturbing to RJB, since the price levels in the contracts were greatly in excess of the equivalent prices of internationally-traded steam coal imports. Although the calculation cannot be precise, the scale of the price benefit in the contracts can be seen in Table 7.6.

For the three years 1995/6 to 1997/8, the differential between RJB contract and import prices was about £8 per tonne, so that with an

Table 7.6:   RJB Contract Prices and Imported Prices. 1995/6 to 1997/8. £/GJ

	Contract prices [1] (Sept. 92 £)	Contract prices [2] (1997/8 £)	Equivalent Import price [3] (1997/8 £)	Differential
1995/6	1.41	1.59	1.32	0.27
1996/7	1.37	1.55	1.21	0.34
1997/8	1.33	1.50	1.17	0.33

(1)   Contract prices are pithead
(2)   1997/8 money values, using GDP deflator
(3)   'Equivalent import price' calculated by taking average price of power station coal
       imports into EU (on a calendar year basis: *IEA Coal Information* 1998 Table 2.3) and
       adding 15p/GJ for the *additional* transport cost for imports against RJB coal at
       'inland' power stations.

average contract tonnage of 29m. tonnes, the price differential was equivalent to about £230m. per annum. Yet in the calendar years 1995 to 1997, RJB's pre-tax profits were about £180m. per annum. Although it would be unwarranted to take this calculation as a precise measure of the financial benefit which RJB derived from the contracts, since there would have been no realistic possibility (because of port-handling and other transport constraints) that the *whole* of the contract tonnage (i.e. nearly 30m. tonnes a year) could have been replaced quickly by imports. Nevertheless, it was clear that the profitability of RJB Mining (and particularly its deep mines) depended overwhelmingly on the contract prices, which had resulted from the policy decisions in 1989/90 and 1992/3 that support for the coal industry should be embedded in the coal prices paid by the electricity generators, rather than as direct subsidies to cover the difference between production costs and world prices.

Thus, by early 1997, it was becoming evident that RJB would be facing the prospect of significant reductions in both volume and price when the contracts expired in March 1998. Other UK coal producers faced the same problem to a greater or lesser degree. The difficulties were compounded by the great imbalance of market power in favour of the major generators, as against the coal industry: while RJB Mining would continue to be dependent on the major generators for 80–90 per cent of its sales (with no realistic bulk alternative market), the generators had the options of imports and/or additional use of natural gas.

There were two further reasons why the generators would not wish to see a continuation of coal contract conditions favourable to RJB beyond March 1998. First, at the time the contracts were negotiated in 1993, with the help of government, it was acknowledged by BC and the other parties concerned that the high prices and fixed volumes had been agreed only because the generators had been able to secure 'back-to-back' contracts with the RECs supported by the latter's monopoly franchises (mainly in the domestic market). In these circumstances, the planned ending of the RECs' franchises with the coming of domestic supply competition in 1998 posed a serious threat to the coal industry. Second, not only did the coal contracts embody a large price premium to RJB Mining (compared with imported coal prices), but the associated coal-backed contracts between the generators and the RECs provided for a substantial profit margin for the generators. As the MMC said in its April 1996 reports on the major generators' bids for two RECs: 'The effect of the coal-backed contracts was to offset operating losses incurred by non-coal contract-related generation and, until 1998, will account for a substantial proportion of National Power's gross margin' (Para 4.28 of the MMC Report 'National Power PLC and Southern Electric PLC' Cm. 3230: Similar observations apply to PowerGen Report). The removal of this benefit in March 1998 would intensify the wish of both generators to see much lower coal prices, particularly as greater emphasis was being placed on the development of competition in generation, as instanced by the Regulator's requirement for divestment of 6GW of coal-fired plant from National Power and PowerGen to Eastern Group in 1996.

In addition to these impending commercial difficulties, the coal industry was faced with more rigorous environmental regulation. On 26 March 1996, Her Majesty's Inspectorate of Pollution (HMIP – shortly to become incorporated in the new Environmental Agency) announced the conclusion of its review, under the Environmental Protection Act 1990, of the authorisation of emissions at power stations in England and Wales (HMIP News Release HM 396, 26 March 1996).

The revised limits represented 'Best Avoidable Techniques Not Entailing Excessive Cost' (BATNEEC) for individual existing plants, together with overall 'portfolio release limits' (colloquially known as 'bubbles') for each generator to allow reasonable flexibility in the loading and operation of individual stations. The main emphasis was on SO_2, where the effect of the new limits would be to achieve reductions in emissions (compared with the actual level in 1991) of 36 per cent in 1997, 53 per cent in 1999, 79 per cent in 2001 and 85 per cent in 2005. In absolute terms, the reduction in emissions was very large. (Table 7.7)

In terms of the effects on UK coal, the new limits in the early years were largely irrelevant, as coal burn was constrained by the impact of CCGTs already operating and committed so that there was likely to be substantial 'headroom' against the SO_2 emission limits. From 2001, however, the new emission limits appeared to represent a serious threat to the coal industry. Even with a high level of continued operation of the 6GW of coal-fired plant fitted with FGD and given an average sulphur content of 1.5 per cent for UK coal supplied to non-FGD stations, the maximum tonnage of UK coal that could be supplied to generators in England and Wales would be about 30m. tonnes in 2001 and 25m. tonnes in 2005. Only by switching to low-sulphur coal imports would coal-fired plant be able to achieve higher utilisation.

At the same time as the new sulphur limits were promulgated, HMIP also published a report: 'Power Generation: A Review of the Way Forward'. Although this was not a policy document, HMIP said that 'the results of this work will be used in the formation of Government Policy'. The main conclusions of the report which were relevant to coal were distinctly discouraging, namely that there was limited scope to improve the environmental performance of older coal-fired plant at economic cost. More particularly, there was no significant environmental advantage, nor would it be economic, to fit FGD to old coal plant that was mainly being operated at intermediate or peak load. Also, although environmentally acceptable technologies using coal were evolving, these were not yet economic: the capital cost of coal-based integrated gasification combined cycle (IGCC) was likely to be at least twice that of gas-fired CCGT.

However, notwithstanding the unfavourable omens on the commercial and environmental factors which would bear upon the longer-term prospects of the UK coal industry, RJB Mining continued to take an optimistic view of the future. In September 1996, the Company had announced that it was preparing proposals to access up to 450m. tonnes of top quality reserves at Witham near Newark 'in the largest coal bonanza since the development of Yorkshire's Selby coalfield 20

Table 7.7:   Overall SO_2 Emission Limits for National Power, PowerGen and Plant Divested to Eastern Group. Kilotonnes SO_2

1995/6 (Actual)	1422
Limits	
1997	1500
1999	1100
2001	500
2005	365

years ago'. More specifically, the Company was developing plans for a new mine capable of producing 3m. tonnes a year for 50 years 'which will give us access to long-term cost-effective energy supplies for generations to come ... Proven and probable underground reserves at RJB pits were increased by 138m. tonnes last year [and] are now in excess of 500 million. The Witham reserves effectively almost double the base of proven and probable reserves available to RJB'. The Company also reported that reserves of 72m. tonnes were available at the mothballed Thorne colliery, and that RJB 'will be developing mining plans over the course of the next few years' (RJB press release 10 September 1996). These prospects, together with an agreement with Texaco to begin feasibility studies on a clean coal (IGCC) plant adjacent to Kellingley colliery, enabled the Company to say that it was 'taking a long-term and strategic view of the business' (*RJB Mining: Annual Report 1996*, p.2).

However, this optimistic outlook was not shared by the Coalfield Communities Campaign (CCC), an organisation of local authorities with interests in the coal industry. In January 1997, they published a report: 'A Fair Deal for Coal – A Fair Deal for Britain' which was expressly sub-titled as 'a strategy for coal for a new Labour government'. The CCC took the view that only a radical change in government policy could avert disaster for the deep-mining industry: 'If Labour does nothing, more than half Britain's remaining 29 collieries could close in the next five years.' Such an outcome was, in CCC's view, likely in view of continuing plans to build more gas-fired plant, the new SO_2 limits for power stations which 'set impossible targets for a decent sized coal industry', and the fact that, when the current contracts with the generators expired in March 1998 'it is highly unlikely that anything similar to the five-year 30m. tonnes per annum deal currently in operation will be offered to the mining industry'. But without such long-term contracts, it would be difficult to justify major investments in new reserves, and RJB's proposal for a new mine at Witham 'would make little sense against this backdrop. Yet without this long-term development the industry will just wither away.'

The CCC considered that such an outcome would be unwarranted, on the grounds that coal-fired power stations using UK coal 'produce the cheapest electricity of any fuel – cheaper than nuclear stations, cheaper than new gas-fired stations', and that the continuing operation of these nuclear and gas-fired plants at the expense of coal represented continuing distortions of the electricity market. On the other hand, the CCC did not consider that a collapse in deep-mined output was inevitable, given a change of government at the forthcoming General

Election: 'with firm government action and more effective regulation, Labour could produce a balanced approach to energy that gives a fair deal for coal and a fair deal for Britain'. The approach was, for the most part, based on the view that once market distortions had been removed, and a 'level playing field' established, UK deep-mined coal would be able to hold its own.

The CCC proposed that an incoming Labour government should institute a full investigation into Magnox nuclear stations' running costs with a view to establishing a case for early closure; announce that it would not normally expect to offer any further Section 36 licences to gas-fired stations, and should curtail the use of gas by dual-firing in coal stations; require the electricity regulator to undertake a fresh inquiry into existing long-term contracts for gas supplies to CCGTs and for the sale of their electricity to RECs (the so-called 'sweetheart deals'), given that there was a prima facie case that they restricted competition in electricity generation; transform the Fossil Fuel Levy (FFL) into a 'clean fuel levy' to help finance clean technologies, including 'clean coal'; encourage the generators to invest in clean coal, and ensure that FGD stations were fully utilised. In addition, the CCC proposed that Labour should require the Environmental Agency to review the new targets for SO_2 emissions in 2001 and beyond 'with a view to setting new limits that take fuller account of the implications for the coal and electricity industries'. To ensure that the above measures flowed through to the benefit of deep mines, CCC recommended that 'Labour should introduce new planning guidance for opencast mining that sets stricter controls and expectations.'

Perhaps the most revealing proposal concerned the issue of replacement coal contracts. CCC says:

> It may be to the electricity generators' benefit in the short-term to put pressure on British mining companies to sign short-term contracts and reduce prices. But this could put at risk the long-term viability of the [coal] industry and risk the nation losing access to precious coal reserves. The coal and electricity industries are now in private hands, but it would be wrong for the government to act as a mere bystander to the negotiation of the post-March 1998 contracts ... Labour should ensure that in negotiating new coal contracts, the electricity generators are required to take full account of the wider national interest.

CCC envisaged that if these policies and measures were implemented, the deep-mined output could be sustained at 35–40m. tonnes. per annum to at least the turn of the century, to the benefit of the electricity consumer and the country.

The CCC document was in many ways a recapitulation of the

arguments which the coal industry and its supporters had put to the Coal Review in 1992/3, and which had then been largely disregarded. But the coal industry still felt understandably aggrieved at the way in which the Major government had permitted – indeed, encouraged – the 'dash for gas' in power generation, notwithstanding the fact that the total costs (including capital charges) of new CCGTs were generally higher than the avoidable costs of generation by existing coal-fired stations, using UK coal. The validity of this complaint perhaps led the coal interest to place too much faith in the removal of 'market distortions' as a solution for the future (as distinct from an explanation of the past). The CCC proposals also displayed an unstable blend of pro-competitive and protectionist measures. Nevertheless, the CCC document was a forcefully-argued case, with which an important part of traditional Labour Party support could readily identify.

New Coal Politics under New Labour

By the time of the fall of the Major government, and the accession of Tony Blair's New Labour administration in May 1997, it was becoming increasingly evident that the prosperity which the privatised coal industry had enjoyed since 1995, could not be sustained beyond 1997/8. With unchanged policies, once the contracts with the generators expired in March 1998, the industry would be exposed to the threat of lower sales volumes and lower prices; and would once again become a high-risk, declining business.

We have seen how the profits and stability of the privatised coal industry, and in particular of RJB Mining, had been due in large measure to the large-scale closures and redundancies carried out by British Coal before privatisation, the largely 'clean sheet' of past liabilities on privatisation, and subsequent consolidation of the product-ivity and cost gains previously made by British Coal. But, above all, the industry had benefited from the government-brokered contracts with the generators inherited from British Coal. In a very real sense, the expiry of these contracts in March 1998 would constitute the final (delayed) stage in the process of privatising the coal industry. The irony was that it would happen under a Labour government.

The first reaction of most commentators (including the author) was that the change of government would make little difference to policy on coal. The number of miners was falling towards 11,000, compared to 235,000 at the time of the last Labour government in 1979; New Labour had been a major (if inadvertent) beneficiary of the defeat of the Great Strike in 1985 in reducing the power of Arthur Scargill and of trade union militancy generally within the Party; and New Labour was keen to show that its policies were pro-competitive and designed to promote the modernisation of the UK economy and environmental protection, rather than the protection of 'sunset' industries, such as coal. More specifically, the new government adopted for the energy sector the Thatcherite framework of private companies operating in competitive markets. Even though the Coalfield Communities Campaign had raised the banner: 'The Coal Industry in Danger', there was little indication in the period immediately following the Election, that the new government was seriously contemplating inter-vention in the electricity market, or in environmental regulation, to help the coal industry.

This placed Richard Budge, the Chief Executive of RJB Mining, in something of a dilemma. The Company was still performing well, and he had no interest in damaging the confidence of the City or the workforce by premature warnings about future prospects. At the same time, if the judgement was that only government intervention could secure the Company's prosperity in 1998 and beyond, the government had to be convinced, and this would involve gathering wider political support. In turn, this would involve painting a stark picture of the consequences of government inaction: and this could not be done other than very publicly. By June 1997, Richard Budge was beginning openly to advocate some of the proposals which had been put forward by the Coalfield Communities Campaign in their January document. Speaking at an exhibition of mining machinery manufacturers, he urged 'our political leaders to invest now in [clean] coal burning technology and make sure Britain can make good and efficient use of its most abundant and only secure and sustainable energy resource – COAL. The Government's first step in achieving that objective should be to impose a moratorium on the granting of any further licences for new gas-fired stations until the social and economic issues ... have been carefully considered' (RJB Press Release 4.6.97). Budge repeated this message in his address to the UDM Conference on 12 June: 'The Labour Party has always been a good supporter of coal in opposition, and we await its action now in power to put the legislation in place that will provide a balanced energy mix thus reducing the over-dependence on gas power' (RJB Press Release). In August, RJB put forward a 'Four Point Plan' (RJB Press Release 3.8.97) as part of the consultation with the Electricity Regulator on 'Price Restraints in the Competitive Electricity Market from 1998', in which the Company proposed extending the existing coal contracts with the generators, but with reducing tonnages linked to a phased introduction of open competition in the domestic electricity market; extending the generators' existing coal-backed electricity contracts with the RECs, with a similarly declining volume and real price reductions; reducing by 11.5 per cent the price of coal from April 1998, thereby 'further reducing the gap between the cost of British-mined coal with that available on the international market'. These proposals indicate that RJB was prepared to recognise that some reduction in both volume and price was inevitable, but was seeking to mitigate the impact likely to arise from free negotiations with the generators. However, the RJB proposals found no favour with the electricity regulator (Professor Littlechild), and were not pursued further.

Already doubts about future market conditions were affecting RJB's attitude to long-term developments. In June 1997, RJB received a conditional licence from the Coal Authority to work coal at the Witham prospect. As the necessary exploration had largely been completed by British Coal prior to privatisation, the next stage involved 'environmental and other studies necessary for a planning application to be made early next year'. However, at a press conference (*Times* 5.6.97), the Company indicated that the project, which would cost £300m. and would take a decade to complete, would depend on the government showing commitment to coal, and that the project was unlikely to proceed for several years, given current market uncertainties. In July 1997, the Company announced that its major development at Calverton, designed to increase the life of the colliery from four years to over thirty years, would be discontinued until the outcome of the future contract negotiations was clear. Of even greater symbolic significance, in August 1997 RJB announced the closure of Asfordby new mine, due to exceptional geological difficulties, and heavy financial losses. Asfordby, which began production in 1995, had been the last new mine initiated by British Coal under 'Plan for Coal', at a capital cost of £360m. (including £40m. by RJB). Its closure marked the end of a chapter. Taken in isolation, the closure of one mine would not have been overly significant, particularly as the reasons related mainly to local geological difficulties. But the Asfordby closure was in addition to earlier closures by RJB of Point of Ayr and Bilsthorpe collieries, which were judged to have reached the end of their economic reserves. Together with the closure of Coventry, Trentham and Markham Main collieries following the collapse of Coal Investments in early 1996, and of Monktonhall in Scotland, some 8m. tonnes per annum of potential deep-mined coal inherited from British Coal had been lost. In any extractive industry, such closures were to be expected. The real issue was whether expectations were sufficiently favourable to allow large-scale investment in replacement capacity. In 1997, this was manifestly not the case, notwithstanding the RJB announcement on the Witham prospect.

However, for the political process once again to be involved with the future of the industry, there needed to be a clear perception, not of inadequate long-term investment, but of impending crisis.

The announcement of the Asfordby closure in August 1997 had caused the coal industry's supporters' alarm bells to ring. Analytic articles on RJB's problems began to appear in the broadsheets (*Sunday Times* 17.8.97, *Times* 19.8.97). Gerry Mousley, Director-General of the Confederation of UK Coal Producers (which was dominated by RJB Mining) was quoted as saying that he had written to the Prime Minister to press for action, adding 'we have had 100 days of New Labour and no indications of support for the British coal industry'. (*Guardian* 19.8.97)

By September 1997 – only six months away from the end of the major coal contracts – the portents were looking increasingly ominous. Coal stocks were 5.5m. tonnes higher than in September 1996, and cumulative coal consumption for electricity generation had fallen by 8m. tonnes compared with the first nine months of 1996. International coal prices were continuing to fall. A 'cliff-edge' crisis for RJB in March 1998 now looked probable. New CCGT commissioning would further reduce coal consumption in England in 1998/9; the generators would have an interest in meeting some of their coal demand by lifting stock; UK coal supplies from sources other than RJB were likely to increase (particularly from Scotland); and, given the low prices available for low-sulphur coal from abroad, imports could well increase. The net effect would be to squeeze the market for RJB, which would effectively become the residual supplier to English generators. The arithmetic suggested that up to half of RJB's deep-mined output might be put in jeopardy.

Yet still the government seemed disinclined to intervene. In its annual Energy Report (*The Energy Report: Shaping Change* Vol. 1. 1997) published in September 1997, the foreword by the Energy Minister (John Battle), referred in general terms to 'the strategic trade-offs between the environmental advantages of CCGT stations – particularly in terms of carbon and sulphur – and the security and diversity arguments for coal and other black fuels' (1.16): but his summary of 'some broad strands of our policy' contained no measures which might be of help to the UK coal industry. (1.19) The Energy Report also set out succinctly the major problems facing the industry

– the effects of CCGTs, contract renegotiation at a time of low international coal prices, and environmental pressures – but suggested no specific remedies, other than support for complaints about subsidies granted to German and Spanish coal industries. Indeed, the Report drew particular attention to the environmental disadvantages of coal, in the light of the government's manifesto commitment to reduce CO_2 emissions by 20 per cent below 1990 levels by 2010 (10.14–10.17).

At the time, John Battle believed that there was no reason why he should refuse Section 36 consents for new CCGTs: and in August and October 1997, he approved three CCGT plants with a total capacity of 1.6GW, even though opposition to such additional consents was becoming *the* focus issue for the coal industry's supporters within the Labour Party. These consents were also noted by the City, which was beginning to change its mind about RJB, after the euphoria of 1995 and 1996 when RJB shares hugely out-performed the market. Two reports by Dresdner Kleinwort Benson (DKB) illustrate this change of perspective. The report in April 1997 said that 'We believe that a Labour Government would be helpful to RJB.' But, by the end of October, DKB were pointing to a serious situation ahead: 'The scale of the volume collapse and price reduction [facing the Company] means that the profit outlook is profoundly affected'; 'Financially, RJB can survive, but will clearly be a shadow of its former self as we move into the next millennium.' Among the adverse developments since the previous report, 'Most worrying has been the total lack of support offered by the new Labour Government. Not only has it further encouraged gas, [and] approved additional CCGT plant, but is also prioritising its environmental commitments over and above the well-being of the coal industry.' And the *Times Business News* observed (14.10.97): 'Whatever new Labour said in Opposition, however, it does not recognise any debts to the miners. They are as friendless as they were under the Tories'. By the end of October 1997, RJB shares stood at 183p compared with 435p. in early May (when New Labour came to power) and their peak of 625p during 1996.

But the re-entry of politics into the affairs of the coal industry was due not to adverse sentiment in the City; nor to public opinion, which was largely oblivious as to what was happening; nor to criticism by the Conservative Opposition, which had no leverage (given the Blair government's huge majority) and in any case was ill-equipped by recent history to pose as the friend of the coal industry. What mattered was increasing anxiety and criticism from coal industry supporters within the Labour Party. These supporters had been mortified by the unwillingness of ministers to meet a coal industry delegation, led by

Richard Budge, at Downing Street on the fifth anniversary of British Coal's closure announcement which had triggered the coal crisis of October 1992.

John Battle continued to resist government intervention in the contract renegotiation. In a statement, he said 'the power generators are private businesses. I have no say at the negotiating table. I don't know why Mr Budge thinks I have', and any subsidy targeted at RJB 'would be grossly unfair to all other companies in negotiations'. (Reported in *Guardian* 22.10.97). This attitude was unacceptable to the coal industry supporters in the Labour Party. *The Times* reported that 'the first showdown between new and old Labour is imminent and the battleground is Britain's coalfields ... The pressure coming from the heartland of old Labour goes straight to the core of the divide between the Government and its traditional party roots.' A meeting orchestrated by the Coalfield Communities Campaign at Kellingly colliery in Yorkshire, and co-ordinated by John Grogan, Labour MP for Selby, proposed a backbench alliance to defend the industry, which, it was thought, could attract the support of as many as 80 Labour MPs. (*Times* 24.10.97) Always sensitive to such political under-currents, at the end of October, the Trade and Industry Select Committee (TISC), with its Labour majority, announced that it had decided to undertake an urgent inquiry into the future of the coal industry, and would request John Battle to give evidence on 3 December to explain the government's stance of non-intervention.

Subsequent events increased the political pressure on the government to be seen to be doing something to help coal. By the end of October 1997 RJB had secured deals with National Power and Eastern Group for firm tonnages in aggregate of 12m. tonnes in 1998/9 and 9m. tonnes in the following two years, at prices some £8 per tonne below those in the current contracts. PowerGen did not settle, claiming that it could probably manage without any RJB coal in 1998/9. The journal *Coal UK* called this a 'contracts disaster for RJB'. It appeared that, even if PowerGen were eventually to settle for 3 or 4m. tonnes, the total contract tonnage in 1998/9 would be only about half the annual level provided for under the inherited 'British Coal' contracts. The cliff-edge scenario for March 1998 now looked very likely: a view which the DTI appeared to endorse. As part of a statistical exercise for the European Commission, DTI indicated that the number of miners directly employed in deep mines would fall by nearly 5,000 by the end of 1998. The DTI return was leaked, and predictably caused indignation among coal industry supporters, and increased the pressure on John Battle in particular. (*Times* 21.11.97) This indignation was further increased by another leak of a DTI internal paper entitled 'Coal Press Strategies' which purported to say that 'there is an expectation that deep-mined coal will effectively end within 10 years unless the Government intervenes to protect it by changing its environmental and energy policies', and that ministers 'must be able to spell out why direct government action is not feasible'. (*Independent on Sunday* 23.11.97)

Battle came in for further attack from Labour backbenchers during a Commons debate on the coal industry on 26 November, but, although he indicated that he was actively considering matters of interest to the coal industry (for example 'working hard to find ways to support Clean Coal Technology'), he was short on specifics, and maintained his previous position on the two particular points where his critics wished to see movement: namely, towards a moratorium on new CCGTs, and direct government intervention in the contract negotiations.

New Labour Intervenes to 'Save the Coal Industry'

As the date for Battle's appearance before the Select Committee became imminent, there was an appearance of confusion in the government's camp. There were reports of initiatives from No. 10 Downing Street to ensure the coal industry's future over the next five to seven years, and of the involvement of Richard Caborn (Minister for the Regions, and previous Chairman of TISC) and, by implication, of John Prescott, the Deputy Prime Minister. On 3 December, less than an hour before John Battle was due to appear before the Select Committee, the Prime Minister told the House of Commons that the government had decided to halt the building of further gas-fired power stations. Battle spelt out the details to the Select Committee. Only two days before, he had received a letter, of almost providential convenience, from the Chairman of the National Grid, who pointed to the increasing dependence of the electricity system on gas-fired generation, and the problems this might raise. It was this which enabled Battle to say to the Select Committee that 'I am announcing a review to look at how the issues of security of supply and fuel diversity should be addressed in considering applications for power station developments, and I propose therefore to defer decisions on outstanding applications until the work of that review has been complete ...' (TISC Report 'Coal ' HC 404-II para. 190).

While the decision on a moratorium for new CCGTs was greatly

welcomed by the coal industry and its supporters, it was recognised
that it would take time to have effect. The immediate problem was
RJB's contract negotiations. On this also, the Prime Minister duly
obliged. On 10 December, he told the House of Commons that a deal,
brokered by the Paymaster General, Geoffrey Robinson, had been
made between RJB and the three major generators for the latter to
continue to take coal in the June quarter 1998. To cheers from the
Labour benches, Mr Blair said that this 'will allow the UK deepmine
coal industry to continue production at present levels without
immediate redundancies or pit closures', and that the government
would use the six months delay to carry out a wide-ranging review of
Britain's energy requirements (Hansard, House of Commons 10.12.97).
On 22 December, Margaret Beckett, President of the Board of Trade,
announced the terms of reference of the government's review of energy
sources for power generation which would 'take as its starting point
trends in energy sources for power generation, especially the growing
dependence on gas, taking into account the energy policy objective of
secure, diverse and sustainable supplies of energy at competitive prices
and, in particular, the role of coal' (author's italics).

This was to be the centrepiece of the government's energy policy
review, but other initiatives were also being pursued; a review of the
Electricity Pool (earlier announced by John Battle in October 1997) to
ensure that there was a level playing field for coal; examination of
some aspects of gas supply contracts to CCGTs; review of the
opportunities for clean coal technology; pressure for reform of the
subsidy arrangements for the Spanish and German coal industries so
as to create opportunities for UK exports; and possible curbs on
opencast production. And Michael Meacher, the Environment Minister
told a meeting of coal industry supporters attended by John Battle,
that his department would not allow any pit to close as a result of
tightening emission limits (CUK 72/7). The sight of an apparent
competition among ministers, (including Peter Mandelson, the high
priest of New Labour) to see who was the most active in the pursuit
of salvation for the coal industry, called into question the government's
executive competence. The standing of the DTI and its Energy
Minister, John Battle, had been damaged, and it was no longer clear
who was in charge of energy policy in general, or 'saving the coal
industry' in particular.

The overall impression given by all this actual and promised activity
was that this was a U-turn driven solely by the government's wish to
placate the coal lobby within its ranks (especially at a time when it had
offended the Left by proposals to reduce social security benefits to

some single mothers). On the other hand, there were widespread doubts among the coal industry's supporters, whether the government had done anything more than postpone the coal industry's problems from March to June 1998. As Bill Flanagan, Chairman of the CCC, said: 'Let's not kid ourselves. None of the proposals so far will necessarily give a single job or sell one more tonne of coal next year or the year after that' (*FT* 17.12.97)

For the government, there was to be no early relief from those entanglements. Finding an effective solution to the coal industry's difficulties which would satisfy the coal lobby and the Left, while at the same time being consistent with the government's pro-competitive position and environmental enthusiasm, was a difficult and complex task. Above all, it was very evident that the coal industry sat on the very fault-line which divided Old Labour's sentiments from New Labour's ambitions.

The exchanges which had taken place before the Select Committee on 3 December (together with supporting memoranda) also brought out several other aspects of the coal problem. First, the Confederation of United Kingdom Coal Producers (COALPRO) took as their main proposition that 'the problems that the coal industry is facing in terms of the share of the fuel market are largely a function of anti-competitive practices in the generation market' (COALPRO Memorandum HC 404 II, para. 3.7), and that therefore the appropriate policy was the removal of market distortions to give coal a 'level playing field': the policy of 'fairness not favours'. However, RJB's position was more ambivalent. John Battle reported to TISC on 3 December that he had received a letter on 11 November from RJB headed 'Urgent request for coal subsidy', and added that 'we ought not to be, as a Government, subsidising profitable firms' (Minutes of Evidence HC 404-II.199). On the other hand, in his oral evidence, Richard Budge drew attention to the difficulties RJB faced in maintaining its previously profitable position. Recently, deep-mine costs had averaged 123/124p/GJ with opencast costs just under 100p/GJ, to give an average of 119p/GJ. The Company had been seeking prices of 125p/GJ; but if prices fell below 120p/GJ, there would be very little, if any profit: 'To maintain a strong viable deepmine industry we need at least 10 per cent net profit to maintain the high level of investment required and to cover the investors' risk in mining' (RJB Memorandum, HC 404 II, paras. 3 and 6).

The Coalfield Communities Campaign, in many ways the authentic voice of the coal interest within the Labour Party, after submitting evidence to the Select Committee in line with its previous position, subsequently came out much more strongly in favour of a more overtly interventionist policy. In a document 'A Market for Coal', published in January 1998, they said:

'There has long been widespread support, especially in the Labour Party, for a "balanced" energy policy that recognises that all the major fuels have a role to play. The key point is that if balance is now to be maintained, in order to deliver long-term security of supply, UK coal should be given a guaranteed, minimum market in the electricity sector' (which would be fixed at a level which would avoid any deep-mine closures) and which would be achieved by requiring the RECs collectively to buy a minimum proportion of UK coal-generated power. Moreover, this guaranteed market should 'provide a framework of long-term stability (15–20 years) that would justify investment in opening up new coal reserves to replace those being exhausted.'

The Trade and Industry (Select) Committee did not accept this approach. Its report on Coal (TISC HC 404-I. Published 24.3.98 but issued 7.4.98) was rather low key, particularly considering that its membership had a majority of Labour back-bench MPs. On the central issue, the Committee said 'we do not suggest that a deep-mined coal capability should be preserved at a given level no matter what the environmental or economic cost'. However, it continued,

we regard it as the Government's task to ensure that deep-mined coal does not labour under unfair structural or other disadvantages in the market for power generation or other markets; that it is not excluded from the market by arbitrarily set emission standards; and that the industry is not allowed to perish in the immediate future if new technology offers it genuinely promising longer term prospects. It is also incumbent on Government to recognise the social and economic consequences of contraction of the deep-mined coal industry, and to ensure that the appropriate remedial measures are taken (HC 404-I, para.7)

On the two issues raised when the government had sought to propitiate the coal interest in its announcements of 3 and 10 December, the Committee was inclined to be sceptical, even critical. First, the Committee did not agree with the government's announced restrictions on new gas-fired stations: 'We do not believe that [this] will assist the coal industry in the short run; nor is it necessarily to the nation's advantage to resist the development of new power-station capacity' (HC 404-I, para. 41); and, second, on the activities of the Paymaster General (Geoffrey Robinson), the Committee observed: 'While we recognise the Government's intentions in seeking at least a short-term resolution of the coal crisis in this way, we remain to be convinced that an exception should have been made to the rule that the

Government should not intervene in commercial transactions between private enterprises ... and note that the outcome of the [December] deal between the generators and RJB Mining may well be rather less propitious than was initially indicated' (HC 404-I, para. 52).

This scepticism was well-merited, but Geoffrey Robinson ploughed on in his quest to broker supplementary deals beyond June 1998 (since his temporary measures had involved mainly rescheduling previously agreed tonnages). Just before the issue of the Select Committee report it was said that the government had 'saved deep-mined coal' (*Sunday Times* 5.4.98). In May there were reports that the government was poised to announce a five-year deal which would safeguard most deep mines (*Times* 13.5.98); and in early June, Geoffrey Robinson was said to be putting finishing touches to a complex deal with generators which, together with other policy initiatives, was shortly to be announced by Margaret Beckett, President of the Board of Trade (*FT* 3.6.98). But when, on 25 June 1998, the President of the Board of Trade announced the launch of the Consultation Paper on the Review of Energy Sources for Power Generation, there was no news of government-brokered contracts. Any solution to the coal industry's problems would now have to await the outcome of this Review.

The White Paper on Power Station Fuelling

Following the Consultation period, the White Paper on 'Conclusions of the Review of Energy Sources for Power Generation' (Cm. 4071) was published on 8 October 1998. The government recognised that the Review had been triggered by anxiety about the future of deep mining, but wished to emphasise that any action it took should be consistent with support for competition. Peter Mandelson (who had succeeded Margaret Beckett but with the title of Secretary of State for Trade and Industry) said when introducing the White Paper: 'I am convinced that competitive markets are the best means of stimulating efficiency in industry, of providing consumers with real choice and bringing down prices. They are the cornerstone of our approach to energy and power generation' (DTI Press Release P/98/768). 'The Government's energy policy is not designed to favour any one fuel over another, to favour indigenous sources or to set restrictions on imports' (Cm. 4071, para. 4.21). Thus, as Mandelson said in his statement 'This policy has no guaranteed market share for coal. The coal industry has asked for fairness not favours. That is what the Government's policy offers.'

The scope and content of the Review, and the conclusions in the White Paper, went much wider than the narrow issue of the market share of UK deep mines in fuelling UK power stations. Yet the link was maintained through the use of two interrelated ideas: first, that the rapid and continuing erosion of coal's market share arising from the 'dash for gas' was in large measure the result of market distortions in the generation market in England and Wales, which had not only disadvantaged coal, but also had kept electricity prices higher than they need have been; and, second, that the consequential increasing dependence on gas for power generation, and the associated displacement of coal would, if allowed to continue unchecked, represent a potentially serious loss of diversity, which might well in due course threaten the country's security of supply, especially as 'coal remains important as the main contributor to the diversity and flexibility of UK electricity production into the foreseeable future' (Cm. 4071, para. 2.12).

In this way, safeguarding the future of coal could be shown to be compatible with both New Labour's pro-competitive agenda and with the national interest. Yet, although the political motivation for the Review and the White Paper had been intra-Labour Party pressure to safeguard output from RJB deep mines (which for many readers of the White Paper was regarded as synonymous with 'coal'), most of the analysis and reasoning in the White Paper related to UK power station use of coal, whether or not supplied by UK deep mines.

The government's analysis indicated that the avoidable cost of generation at *existing* coal-fired plant was likely generally to be cheaper than that at *new* CCGTs (particularly at lower load-factors), and that, therefore, the pressure to continue with new building of CCGTs was evidence of market distortions in the generation market in England and Wales. In the government's view these distortions arose both from the operation of the wholesale market (The Electricity Pool), where 'bids' did not properly reflect costs, and from the continuing power of the major generators to set Pool prices at an artificially high level well above the total costs of new CCGTs, thereby continuing to encourage the profitable entry of such new plant, even though generation from such plant was more costly than from existing coal plant. If the 'dash for gas' were to continue on this basis, not only would there be a misallocation of resources, and an unwarranted further displacement of coal in favour of gas, but the dependence of power generation on gas might rise to up to 60 per cent by 2003, with coal's contribution being no more than 10 per cent, 'This would mean a significant loss of diversity' (Cm. 4071, para. 2.14)

The government's proposed remedies for market distortions were fourfold. Firstly, reform of the electricity trading arrangements in England and Wales by replacing the Electricity Pool with a new system more analogous to a commodity trading system; secondly, seeking practical opportunities for further divestment of 8GW of coal-fired plant by National Power and PowerGen in order to remove their ability to set prices at an artificially high level (the government hoped to see at least a 10 per cent fall in wholesale prices); thirdly, pressing ahead with implementation of competition in electricity supply for all customers in order to reinforce the competitive purchase of electricity in the wholesale market; and fourthly, but more controversially, there would be a 'stricter consents policy' for new power stations which would be 'short term' and 'temporary' and would apply only until the identified market distortions had been removed, but under which 'new natural gas-fired generation would normally be inconsistent with the Government's energy policy concerns relating to diversity and security of supply' (Cm. 4071, para. 2.44)

Other measures proposed by government of relevance to the UK coal market included re-examination of the terms of trade which had led to the French interconnector being used at full capacity for imports from France (Cm. 4071, para. 2.33); and 'working with the European Commission and our European partners to minimise the distortions caused by state aid in the coal mining sector within Europe' (para. 2.35)

So far as the environmental threats to coal were concerned, the government stated that the future role of coal 'is likely to become dependent at some point in the future on the use of clean coal technology' (Cm. 4071, para. 9.27), but 'the DTI does not feel that funding the construction of currently available clean coal plant ... whether by way of a direct grant or some kind of electricity levy, would constitute value for money at present' (para. 9.30). (This approach to clean coal technology was outlined in more detail in the government's Energy Paper 67 'Cleaner Coal Technologies', published in April 1999). But otherwise, the Government sought some mitigation of coal's environmental difficulties. All major coal-fired generators would be encouraged to have at least one FGD-equipped station (Cm. 4071, para. 9.39). (This requirement seemed to have prompted the early announcement of Eastern to fit FGD at West Burton, and subsequent provisional undertakings by new owners to fit FGD to plant divested by PowerGen). The government also indicated that, although the previously promulgated SO_2 limits for 2005 would be adhered to, there would be some relaxation of the phasing of these limits in the interim; and that also more onerous proposals put forward by the Environment Agency in January 1998 would not be proceeded with (paras. 9.45-9.47).

Such an approach could reasonably be regarded as compatible with the UK's international commitments on SO_2 emission reduction. However, the government's attempts to argue that its policy of favouring coal at the expense of gas was compatible with its policies on Climate Change were less convincing. Indeed, the White Paper admitted that it was self-evident that the stricter consents policy [on new CCGTs] would tend to slow the decline in CO_2 emission levels in the short term, but took refuge in the uncertainties in all the factors which might work in the other direction. The government's first consultation paper on the 'UK Climate Change Programme', issued in October 1998 (shortly after the White Paper on power station fuelling) also maintained that: 'Our policy proposals ... are consistent with a continued decline in carbon emissions from the electricity supply industry towards our climate change targets' (para 46). The environment ministers could not bring themselves to say that meeting these targets would be made more difficult unless there were further reductions in coal use.

How far, therefore, should the White Paper (Cm. 4071) be seen as an act of effective political management of the coal industry's problems? The benefits of Pool Reform to coal demand were speculative, since as 'it would appear that in most circumstances existing [or committed] gas plant is cheaper to operate than existing coal plant' (Cm. 4071, para. 5.37), coal was unlikely to benefit significantly from a more cost-reflective wholesale market (unless there was a large increase in gas prices *relative* to coal prices); and there was no certainty that, even if divested coal-plant operated at higher load-factors than previously, plant divestment would result in a significant *net* increase in power station coal consumption. On the other hand, the (temporary) restrictions on the licensing of new CCGTs, the expected fall in wholesale electricity prices, arising from greater competition in the power generation market, and the encouragement of further FGD seemed likely to reduce the impact of gas-fired generation on the coal market in the UK over the period 2000–2005. Thus, the overall effect of the Review appeared likely to be a UK coal market after the year 2000 somewhat larger than it would otherwise have been (albeit smaller than in 1998, due to the impact of new CCGTs already under construction or with full planning consents)

However, the link between these developments and the political objective of 'saving the UK *deep-mined* coal industry' looked very tenuous. In the first place, the White Paper established no *necessary* connection between a somewhat higher level of UK coal demand (than would otherwise have been the case) and the maintenance of UK deep-mined output levels. 'Whether or not UK coal is chosen for use by the generators, in preference to coal from other sources, must be a matter for the commercial decision of those generators and not the Government' (Cm. 4071, para. 8.18). Further, the White Paper did not seek to argue a *necessary* connection between the strategic arguments in favour of limiting the switch from coal to gas, and the case for more UK coal: 'While UK coal is available and the generators continue to choose it, UK coal contributes to our energy diversity and security. If the generators change their preferences or available economic UK coal production falls in the future then international coal will play a greater role, provided that the power stations which burn coal remain available. International coal can therefore provide a useful element of diversity and security to UK electricity supplies' (Cm. 4071, para. 8.25). Moreover, to the extent to which the government's measures resulted in more intensive competition between coal-fired generators (with falling wholesale electricity prices), there would be even greater incentives for the generators to seek lower price coal supplies.

The potential benefits to UK deep-mining of the government's initiatives to reform the coal subsidy rules in Europe were also likely to be indirect and to take some time since the current regime was not due to finish until the expiry of the ECSC Treaty in 2002. In the interim, the German and Spanish governments would be unlikely to agree to an accelerated rundown of their own heavily subsidised coal industries in order to give opportunities to RJB. In any case, so far as the UK deep-mine coal industry was concerned, the problem created by subsidies in Europe was neither direct imports of German or Spanish steam coal, nor the exclusion of UK imports by restricted access (in 1998 there were over 100m. tonnes of steam coal imports into Western Europe, including 29m. tonnes into Germany and Spain). Rather, the problem was economic. A combination of low international prices, and relatively high transport costs, made it uneconomic for virtually any English deep-mined output to be exported to continental Europe. Clearly, if most of the German and Spanish output could somehow be made to disappear by the removal of subsidies, then this might (at least in theory) increase the level of internationally-traded coal prices, and so improve the economics of UK deep-mined coal. But such an outcome was unlikely to materialise, at least not for the foreseeable future.

Thus, the government's proposals in the White Paper, in terms of the benefits to the sales volume of UK deep-mined coal, were either uncertain, or unlikely to materialise in less than two or three years, or both. Yet in 1998 for RJB Mining the problem of finding a market at acceptable prices for all its deep-mined output was immediate. At the time, it was not clear why the publication of the White Paper should lead the major generators to contract for additional tonnages of RJB coal from 1998, in preference to cheaper imports.

The Problem of Opencast Coal Production

Most of the agonising over coal policy by the New Labour government in 1997 and 1998 had concerned the future of the remaining deep mines, particularly those of RJB Mining. Indeed, but for the expectation of significant deep-mined closures in 1998, it is doubtful whether the 'Review of Energy Sources for Power Generation' would have taken place, and certainly not in the highly public way which occurred.

Yet there were also unresolved issues on opencast production. Twenty years before, the proportion of opencast output in total UK coal production had been typically about 10 per cent, but due to the subsequent disproportionate decline in deep-mined output, by 1999, opencast represented 40 per cent of UK coal production. There appeared to be sufficient identified shallow coal deposits to sustain opencast production in the range 15–20m. tonnes a year for the foreseeable future. The problem lay not in the reserves, nor (generally) in the economics, since opencast coal was normally competitive in cost. The main issue was the difficulty of obtaining planning permission for new sites in the face of often fierce local environmental opposition.

The average life of sites had fallen since privatisation, and by March 2000, the remaining reserves in sites with operating licences had fallen to 25m. tonnes, equivalent to less than two years' production, even though the reserves in conditional licences (but without planning permission) amounted to 230m. tonnes (*Coal Authority: Annual Report 1999/00*). Thus a large number of new sites had to be approved by the planning system every year to maintain total output. In the first three years after privatisation, this was broadly the case, although with some differences between regions. (Table 7.8). In South Wales, most of the output was of specialist quality – anthracite and other naturally smokeless fuel for the domestic market and low volatile power station coal for Aberthaw power station; in Scotland, output was mainly of steam coal for power stations, and new sites were generally easier to obtain; but in England, permission for new sites to produce power station coal had become increasingly difficult, with most of the reduction in output falling on RJB Mining, a trend which continued in 1998/9 and 1999/00.

Table 7.8:   Opencast Output. 1995/6 to 1999/00. Million Tonnes

	1995/6	1996/7	1997/8	1998/9	1999/00
England	8.9	8.4	8.1	7.0	6.2
Scotland	5.1	5.4	6.3	6.4	7.2
Wales	2.1	2.3	1.8	1.5	1.5
Total	16.1	16.1	16.2	14.9	14.9

Source: Coal Authority

The Labour Party had always had an ambivalent attitude to opencast coal mining. On the one hand, the generally lower cost of opencast could be used to cross-subsidise deep mines, and so maintain a higher volume of total production; and the low chlorine content of opencast coal could provide a 'quality cross-subsidy' to support high-chlorine deep-mine production, particularly in Yorkshire and the Midlands. On the other hand, opencast was often seen by the Mining Unions and the Local Authorities in the coalfield areas as an undesirable competitor which, given the constraints on the total coal market, should be restricted by government action. This ambivalence had been seen in the CCC Document 'A Fair Deal for Coal – A Fair Deal for Britain' (January 1997) which, as we have already noted, concluded that 'Labour should introduce new planning guidance for opencast coal mining that sets stricter controls and expectations.'

The Labour Party itself issued its '10 Point Plan for opencast' before the May 1997 election which proposed significant restrictions on new sites, designed to 'reduce the reliance on opencast coal as part of an overall energy policy'; and Richard Caborn (the relevant minister within the DETR) made similar noises in government. In October 1998, shortly after the issue of the White Paper on power station fuelling, the government issued a further Draft Consultation document: 'Mineral Planning Guidance Note 3: Coal Mining and Colliery Spoil Disposal', which was subsequently formally issued as MPG 3 (Revised) in March 1999. In this document, the government pointed out that 'although some sites are capable of being well restored, opencast mining can be extremely damaging to the environment and amenity of a locality whilst it is taking place'. There was no national case for opencast: rather, 'the Government believes there should normally be a presumption against development', unless the proposal provided local or community benefits which clearly outweighed the likely environmental impact, and that this judgement was best made by the local communities themselves. Moreover, on the chlorine issue, which had often been regarded by opencast operators (including RJB) as a crucial argument in favour of new sites, the government was dismissive saying that it 'does not regard a need for coal blending [of low chlorine opencast with high chlorine deep-mined coal] as a significant issue for the land use planning system, nor does it see a role for the [planning] system in influencing the operation of the market for coal'. Thus, the government was giving clear signals that it would do little or nothing to support opencast mining in the (probable) event of local opposition, and was prepared to issue formal guidance to the planning system which would increase the chance of such opposition succeeding. Indeed, it was very clear that 'saving the coal industry' did not include the 40 per cent of UK coal output that came from opencast sites.

New Labour Gains Temporary Respite

The New Labour government's intervention in the affairs of the coal industry had not been to safeguard the market prospects for opencast or imported coal, but to protect UK deep mines and their workers. Specifically, we have seen how intervention was prompted by the fears of a 'cliff-edge'-style crisis, with large-scale colliery closures and manpower rundown as soon as RJB's contracts with the major generators expired in March 1998. In an effort to avoid this outcome, the government had secured a rescheduling of coal supplies to maintain RJB's markets to June 1998 while the power stations fuelling review and the other policy initiatives were being pursued.

However, the expected 'cliff-edge' problem of 1998 failed to materialise, as a result of factors unrelated to the government's initiatives to 'save the coal industry'. First, deep-mined output fell by over 5m. tonnes, due mainly to RJB's policy of decreasing the number of working areas underground and concentrating development on fewer districts, as well as major geological difficulties at two mines (*RJB 1998 Annual Report*). Including opencast, total UK coal output declined by 7m. tonnes between 1997 and 1998. Second, against all expectations, UK coal consumption stabilised, and coal use by electricity generators actually increased as a result of delays to the output of new CCGTs, and an interruption of electricity imports over the French inter-connector, due to temporary difficulties at some French nuclear plants. As a result of this unexpected combination of events, a supply/demand balance for UK coal in 1998 was able to be maintained without resource to politically-contentious announcements of large-scale colliery closures (See Table 7.1).

Progress was also made in increasing the forward contract cover for RJB Mining. In December 1998, Eastern Group announced that it had negotiated to take up to 16m. tonnes over five years from April 1998 (in addition to the tonnage negotiated in November 1997), together with a separate 'options' contract from 2003 as and when FGD equipment was installed at West Burton power station. At the same time, PowerGen (which had hitherto failed to negotiate new contracts with RJB) settled for up to 35m. tonnes over five years of supply from April 1998. And this round of contract negotiations was completed in April 1999, when National Power announced agreement to take up to 28m. tonnes of RJB over the following four years, in addition to tonnage agreed in 1997.

In aggregate, these contract volumes (including those negotiated in 1997) looked very satisfactory, totalling 26m. tonnes for 1999/00 and 2000/01, and 17m. tonnes in the following two years. This seemed to indicate that there was a good chance that the contracts would (with suitable phasing) be sufficient to place all the available RJB power station coal until the time of the next General Election. Although the government's official position remained that these contracts had been commercially negotiated between RJB and the major generators, this did not prevent the Energy Minister, John Battle, from declaring (DTI press release 13/4/99): 'It's welcome news that National Power, like other major generators, has recognised the value of UK coal in pro-viding flexible and secure supplies in an increasingly competitive market. This contract is a demonstration that UK coal can compete effectively when it has a fair opportunity to do so. The announcement will be welcomed by mining communities as a vote of confidence in them.'

For a time it appeared that the government's attempts to gain time and to reassure its supporters had been successful. The earlier fears of a 'cliff-edge' collapse of deep-mined output in 1998 had been avoided; and the government had shown its commitment to the coal industry by bringing about (by whatever indirect means) improved contract volumes for RJB Mining, imposing restrictions on the introduction of new gas-fired generating plants, mitigating the impact of environmental regulation on coal, and introducing measures to create a 'level playing field' for coal in electricity generation.

The 'Coal Crisis' Re-emerges: The Slide towards Subsidies

However, New Labour was soon to discover that it had not solved the problem. Certainly, the new additional contracts did not provide a solution for a number of reasons.

First, the supplementary contracts for RJB coal entered into by three major generators owed less to their belated realisation of the benefits of these supplies (at well above international prices) than to the companies' perception of their self-interest in accommodating the government's wishes as far as practicable, and in particular as one of the means of obtaining permission for acquisitions involving vertical integration, which hitherto had been regarded unfavourably by the competition and other regulatory authorities. Thus, PowerGen received clearance for the acquisition of East Midlands Electricity, and National Power for its acquisition of the supply business of Midlands Electricity; while Eastern obtained tacit recognition of the degree of vertical integration which it had already achieved. This whole episode suggested that the new arrangements rested less on commercial foundations than on the ephemeral exercise of the Black Arts of Government.

Second, as was widely reported, the new deals contained substantial tranches of optional tonnage, so that the degree of forward security in volume was subject to considerable doubt, and was likely to taper sharply in the later years, particularly 2001. In this respect, these new contracts were much less favourable to RJB than the 'take-or-pay' contracts inherited from BC. Thus although the short-term outcome was that RJB Mining could reasonably rely on some 20m. tonnes of contracted sales to the generators in both 1999 and 2000 (RJB Interim Results, 1999: Chairman's Statement), the problems of further market contraction had only been postponed.

Third, the level of pithead prices (reported to be in the range of £1.15–£1.20/GJ) was unlikely to provide any significant profit margin for most deep-mined output, even though at the time these prices were 20–30 per cent higher than for equivalent imported coal.

Fourth, and most importantly, given prevailing market conditions, these new contracts provided quite insufficient security in terms of either the duration of firm tonnage commitments or the level of prices, to form the basis for major long-term capital investment in new and replacement deep-mined capacity.

It was not long before the adverse underlying trends once again became evident and politically visible. In the first half of 1999, the decline in UK coal consumption resumed after the pause of 1998, with the power generation sector falling by 18 per cent, and deep-mined output continued to run some 15 per cent below the 1998 level. In January 1999, Midlands Mining (which had taken on the two pits to survive the collapse of Coal Investments in 1995/6), having recently closed its Silverdale mine, announced the forthcoming closure of its remaining colliery, Annesley/Bentinck. In April, RJB announced the closure of Calverton colliery, which until 1997 had been the subject of a major investment project to extend its life; in May, the company issued a profits warning which drew attention to low productivity and adverse geological conditions at the Selby complex (which made up a third of the UK deep-mined output), and in July 1999 announced the effective closure of the North Selby workings; and there followed announcements of the forthcoming closure of Clipstone and Ellington collieries.

There was also rising anxiety that there would be further displacement of RJB deep-mined coal by imports (which in 1999 continued to be 20–30 per cent cheaper than UK coal), when the major coal-fired generators came to exercise their contractual options in respect of supplies for 2001 (which would require decisions in 2000). This risk was reported to be particularly high in the case of the new American owners (AES and Edison Mission Energy) and British Energy, who had acquired 10GW of coal-fired plant sold by National Power and Power Gen, and would seek to defend profitability in an increasingly competitive generation market by cutting their fuel costs.

Against this background, an adjournment debate on the coal industry was held in the House of Commons on 10 November 1999. The main spokesman for the coal interest was Michael Clapham (Labour MP for Barnsley West and Penistone), who stated: 'Government policy was contained in the October 1998 White Paper on Energy, which was supposed to pave the way for a level playing field that would help the coal industry. It is fair to say that, one year on, it is clear that that policy needs reviving if we are to save the coal industry' (Hansard Column 1045). He believed that 'the simplest way to give aid would be to pay the generators the difference between delivered international coal market prices and the UK price, so that the subsidy went to the generators in the short term, until the industry was able to find stability' (Hansard Column 1047). The Energy Minister, Helen Liddell (who had succeeded John Battle) made much of her commitment to the coal industry, and reminded the House that 'I am one of the few women honorary members of the National Union of Mineworkers. I was invited to join by the late Michael McGahey' (NUM Vice-President at the time of the Great Strike). Although she was non-committal, she concluded by saying that 'we shall look seriously at every point that has been raised today with a view to finding a means whereby we can secure the future of the industry' (Hansard Columns 1063-5). And subsequently RJB wrote to the DTI setting out its proposals for financial aid (CUK 96/9).

Certainly, RJB's financial results were much diminished, particularly in the case of its deep mines, where operating profits, which had been £138m. in 1997, fell to £26m. in 1998, and £12m. in 1999, when there was also a write-down of £131m. in the value of colliery assets (See Table 7.3). RJB's share price, which had been 435p when New Labour came to power, had fallen to around 30p by the end of 1999: the City were valuing the company at less than the value of its coal stocks.

For the UK coal industry as a whole, 1999 saw output and sales again falling sharply, continuing the adverse trends seen since privatisation. Between 1995 and 1999, deep-mined output had fallen by over 14m. tonnes (40 per cent), nearly all due to the irreversible closure of capacity; and total sales of UK coal had fallen by over 20m. tonnes (36 per cent), reflecting a similar reduction in coal consumption at power stations as a result of the increased use of gas (Tables 7.1 and 7.4). Of particular significance to the New Labour government, deep-mined output had fallen by 11m. tonnes since its accession to power, and the decline was continuing at an accelerating rate. The sophistications of the power station fuelling review were left behind. The policy of 'fairness not favours' was not enough to satisfy the government's critics from within the Labour party, and the subsidy issue assumed centre stage.

By April 2000, it became clear that neither the Secretary of State (Stephen Byers, who had succeeded Peter Mandelson, and who at the time was under considerable pressure to save jobs at Rover Cars) nor the Energy Minister (Helen Liddell) would wish to be held responsible for further pit closures before the next General Election. On Thursday 17 April 2000, Stephen Byers made a statement on 'Energy Policy' to the House of Commons (Hansard Columns 697-699) in which he announced that the government expected in October to lift the 'stricter consents policy' on new gas-fired power stations; but that because of this, and continuing competition from imported coal, the government was discussing with the European Commission the potential for state aid to the coal industry over a transitional period until the termination in 2002 of the ECSC Treaty (under which rules for coal subsidies were set) for up to a possible total of £100m. over that period. Byers claimed that, on this basis, the government's policy 'will ensure that there is a viable future for the coal industry within a competitive energy market' (column 707).

Although the statement had been short on significant detail, the politics were clear enough. Michael Clapham, who had led the coal debate in the House on 10 November 1999, on behalf of the all-party coalfield communities group welcomed the government's announcement; and the following day RJB Mining announced that it had shelved its plans to close Ellington and Clipstone collieries, and that the future of the Selby supplies to Drax power station could now be secured.

The Inevitability of Continuing Long-term Decline

In the Parliamentary debate on 17 April 2000, Stephen Byers claimed that to subsidise coal was part of a policy to 'support industries that have a real future' (Hansard Column 701). But despite the protestations of ministers, whatever the outcome of the subsidy scheme or future trading conditions, by the year 2000 the inevitability of long-term decline in deep-mine output was evident.

Although the coal industry needed to continue to invest in new and replacement capacity if it was to prevent such decline, by 1999 the industry's capital investment had fallen to its lowest level ever. Only two major capital projects to develop new reserves were in progress – at Daw Mill (RJB) and Longannet (Scottish Coal) – and RJB abandoned its plan for a new mine at Witham, so that its conditional licence from the Coal Authority lapsed (*Energy Report* 1999, DTI, para. 11.8). The IMCL report ('Prospects for Coal Production in England, Scotland and Wales': International Mining Consultants Ltd.), commissioned by the DTI as part of the power station fuelling review, and published in March 1999, had made it clear what the underlying realities were. The report gave an overall assessment of the remaining coal available at *existing* deep-mines in early 1998. (Table 7.9)

But these figures cannot be accepted at face value if economic factors are given proper weight. The IMCL Report included projections of average deep-mine costs: for the four years to 2002, these were £1.21/GJ on an 'operating cost' basis (including depreciation), and £1.12/GJ on a 'cash basis'. With pit head prices for power station coal in the range of £1.15–£1.20/GJ for contracts (and less than this if aligned with 1999 imported prices), not only would it be wholly justified to exclude new mine developments (with total costs of up to £1.60/GJ) from the reserves assessment, but the prospects for major investment even at existing pits would remain very problematic, and the cost pressures arising from tight profit margins would tend to result in still more selective working and lower recovery of existing, accessed

Table 7.9:   Overall Assessment of Coal Available at UK Deep Mines. Early 1998.
Million Tonnes. IMCL Report

	RJB Mining	Other	Total
Reserves [1]	200	20	220
Resources [2]	270	30	300
'Mineral Potential' [3]	90	10	100
Total	560	60	620

Notes: Derived from IMCL Report Table 4.1, and 'rounded'
(1) 'Reserves' are measured mineral resources for which detailed technical and
economic studies have demonstrated that extraction can be justified at the time of
determination and under specified economic conditions
(2) 'Resources' are that proportion of the mineral resource which has not yet been
sufficiently appraised to qualify as reserves
(3) 'Mineral Potential' is coal for which information is limited and extraction is
currently not envisaged.

reserves. (A proposal to develop a new mine at Margam in South
Wales was a very special case, involving coking coal accessed from a
deep opencast site). Therefore, although the above overall assessment
suggested an undiscounted reserves/production ratio of about 30 at
1999 output levels, in reality these numbers required substantial
discount for economic risk. In the absence of very large and sustained
improvements in the *value* of UK coal, it was likely that the greater
part of the 'Mineral Potential' tonnage, and a large part of the
'Resources' would not be developed. (Somewhat arbitrarily, we suggest
discounts of 75 per cent and 50 per cent respectively).

On this basis, and allowing for the subsequent actual or announced
closure of four collieries, and for actual output in 1998 and 1999,
Table 7.10 shows the approximate remaining discounted reserves at
existing deep mines.

(This result is not dissimilar to that using an alternative method
based on the discounted reserves figures in the 1993 Boyd report (See
Table 5.5, Chapter 5), after adjusting for the effect of later closures
and interim output).

A rather more pessimistic assessment was made by the Coal
Authority in its Report for 1999/00. Although it identified the
remaining reserves at currently licensed operating deep mines as 453m.
tonnes at March 2000, the Authority went on to say that experience
had shown that not much more than 50 per cent of available coal was
in fact extracted, as a result of technical, environmental or economic
difficulties. This would suggest a figure of some 230m. tonnes. The

Table 7.10: Remaining Coal Reserves Adjusted for Economic Risk

IMCL Category	M. tonnes
Reserves	170
Resources	140
Mineral Potential	20
Total	330

Source: Author

DTI's *UK Energy Sector Indicators* (published December 1999) implied an even higher discount as 'there were estimated to be approximately 200m. tonnes of economically viable coal reserves at existing mines at the end of March 1999' (p. 23).

Thus, depending on the assessment used, the UK deep-mined industry entered the year 2000 with a reserves/current production ratio of 10/15 years. But, as we have noted before, the remaining reserves were not evenly distributed. Even if existing mines were worked to exhaustion, and even if there were no more premature closures on economic grounds, by the end of 1999 it was clear that total deep-mined output would not exceed 15m. tonnes in 2010, reflecting the loss by then of Selby and some other mines, with further decline thereafter to 10m. tonnes or less.

The Labour government's belated (and, in some quarters, reluctant) intervention had centred on medium-term measures to remove or mitigate market distortions or other constraints that had operated against coal use in power stations, and, at the same time to facilitate higher volume contracts for RJB for the following few years. But this mix of measures did nothing to create the conditions whereby this elderly extractive industry could, realistically, invest sufficiently to replace the progressive exhaustion of its currently accessed coal reserves by the construction of new deep mines or other major developments. Indeed even to have attempted to do so would have required a willingness to impose on the privatised electricity industry massive distortions *in favour* of UK coal (in the form of obligations to take UK coal over a prolonged period at prices greatly in excess of international levels) in ways which would have been wholly incompatible with New Labour's espousal of competitive and liberalised energy markets, and which would have raised considerable difficulties for the public credibility of its Climate Change Programme to reduce greenhouse gas emissions.

The IMCL analysis was available to the government when the

White Paper on Power Station fuelling was drafted, and indeed the main IMCL findings were included in Cm. 4071, albeit in a low-key way (paras. 8.13 and 8.14). It was thus clear to the government at that stage, if not earlier, that there was no realistic possibility of 'saving' the deep-mine industry from further decline, and most ministers appeared to recognise this. As so often in the past, the main policy requirement was to establish a politically acceptable phasing of deep-mine contraction. The New Labour government discovered how difficult this was to achieve (even with the industry much diminished) in the face of Old Labour sentiment, notwithstanding the comprehensive rigour of the government's policy review and its generally 'pro-coal' stance. New Labour was also to find that the Thatcherite policy framework of privatisation and competitive energy markets, which it had adopted, was not capable of delivering a *sustainable* economic deep-mined coal industry.

CHAPTER 8
THATCHERISM AND THE FALL OF COAL: FINAL REFLECTIONS

The Big Question

In the foregoing chapters we have traced the declining fortunes of the UK coal industry (and particularly of its deep mines) since 1979. We now assess the significance of the part played by Thatcherism in this process which, by any standard, is a highly significant episode in British economic history.

Margaret Thatcher came to power in 1979 with strongly-held visceral sentiments on the state of the coal industry, which she was later to describe as symbolising 'everything that was wrong with Britain'. There were two main elements in her viewpoint: first, the need permanently to break the power of the NUM to 'hold the country to ransom'; and, secondly, radically to change what was seen as the inefficient way in which the NCB was operated in the public sector by turning the industry into a viable commercial enterprise (although the policy on ultimate privatisation took some time to evolve).

There is no doubt that these Thatcherite sentiments represented the fundamental coal policy aims of four successive Conservative governments. There was also a wider political significance, not only because these aims derived from the general Thatcherite aversion to trade union power and public ownership, but also because success with coal policy would have a considerable bearing on the success of the Thatcherite enterprise as a whole. In that sense, the new government's aims of coal policy could be said to represent a 'political agenda', representing an important part of what the Conservative governments stood for, in ways which were diametrically opposed by the Labour Party at the time.

As circumstances evolved, the Thatcherite approach proved to have considerable internal coherence, since the two essential elements of policy tended to be mutually reinforcing. The aim of permanently breaking the power of the NUM entailed a large reduction in the dependence of the electricity industry on UK deep-mined coal, which, because of the absence of realistic alternative markets, in turn involved a large reduction in output; and the objective of fully commercial operations under increasingly unfavourable market conditions, pointed not only to lower output, but also to large increases in productivity, and hence very large reductions in colliery manpower.

By the time the period of Conservative rule ended in 1997, this agenda appeared to have been achieved to a degree which could hardly have been envisaged in the early 1980s. Deep-mined output, which had been 109m. tonnes when Margaret Thatcher came to power, had declined to 30m. tonnes. Even more spectacularly, colliery manpower had fallen from 233,000 to less than 12,000. In Yorkshire, which had earlier become the power base of Arthur Scargill's particular brand of trade union militancy, manpower had fallen from 65,000 to little more than 5,000; and in Scotland and South Wales, in many ways the natural allies of Yorkshire within the NUM, collective manpower, which in 1979 was 48,000, by 1997 was less than 2,000. The NUM had become a negligible force in industrial politics, and the dependence of the electricity supply industry on UK deep-mined coal, which had been the fundamental basis of NUM power, had fallen from nearly 70 per cent of power station fuelling in 1979, to around 20 per cent by 1997. Finally, whereas in 1979 the de-nationalisation of NCB/BC would have appeared almost an absurd idea, by the end of 1997 the industry had been operating in the private sector for three years, and was making profits of some £200m a year.

The fact that this outcome seemed to conform so closely to the Thatcherite 'political agenda' has tempted some to conclude that this was a simple case of cause and effect. Those that hold this view see the whole process as driven almost solely by the political will of successive Conservative governments between 1979 and 1997, and implemented by careful (and often secret) planning to bring about the desired result. This view has its attractions both for those on the Right, who see the outcome as a vindication of Thatcherite policies, and for those on the Left, who *by the same token* regard what has happened to the coal industry as the result of a great Tory conspiracy.

This view is mistaken in its simplicity. In the first place, even though it is reasonable to say that Margaret Thatcher's government had a 'political agenda' for coal from the outset, there is no evidence that the new government came into power in 1979 with a clearly thought-out long-term master plan designed to bring about the achievement of its objectives. Still less was there any idea in the early 1980s that these objectives would result in the degree of contraction of the deep-mine industry which eventually occurred. It was in the nature of things impracticable to set out a pre-determined plan of action for what was always likely to be a protracted process, and which in the event covered four consecutive parliaments. Rather, the 'political agenda' for coal represented the terms of reference for coal policy, to be implemented over time as circumstances allowed and opportunities for particular measures presented themselves.

Thus, in seeking to find out why the Thatcherite 'political agenda' for coal appeared to have been achieved so completely and overwhelmingly, we must look beyond the exercise of political will, and search for more complex, interactive and unpredicted reasons.

The Great Strike of 1984/5: A Lucky Victory

Although the first Thatcher government (1979–83) was clear what it would like ultimately to accomplish with coal policy, it initially had little notion how it would achieve these objectives, and made virtually no progress in advancing them. Indeed, it suffered a serious setback in February 1981, when it had to make damaging concessions under the threat of NUM strike action.

The second Thatcher government (1983–7) was rescued from impasse by the defeat of the Great Strike of 1984/85, which broke NUM resistance to the closure of uneconomic collieries and to large-scale manpower rundown. The great political and other benefits which the government derived from the defeat of the Great Strike have led to a widely-held opinion that the government, having made appropriate preparations, mainly by building up coal stocks at power stations, deliberately engineered the outbreak of the strike in March 1984, in order to 'take on the miners'. This is almost certainly wrong, in view of the unpredictably chaotic circumstances in which hostilities broke out in March 1984. And, indeed, any plan deliberately to provoke strike action at that time would have carried great political risks.

On the other hand, there is no doubt that the government wished to break the power of the NUM at some point, and hoped for an opportunity to reverse the humiliation of February 1981. It can be argued that a confrontation between the NUM and the government was virtually inevitable, given that their respective positions were both strongly held and totally irreconcilable. But on this hypothesis, any discussion as to who was 'responsible' for the Great Strike becomes in the last analysis a matter of partisan semantics: who should take the blame, the Immovable Object or the Irresistable Force?

Once the Great Strike had begun, the government was resolute in its determination to prevail, and was willing to underwrite the huge costs involved, particularly the use of fuel oil in place of power station coal. The government also benefited greatly from the fortitude of the working miners and the ingenuity of the managements of the NCB and CEGB in maintaining power station operations. But above all, the defeat of the Great Strike flowed from the attempts of Arthur Scargill and his militant associates in the NUM to launch a national strike without a national ballot, and to prosecute the strike by mass picketing at those collieries opposed to the strike. These actions divided the Union, deprived the NUM of effective support from the wider trade union movement, and alienated large sections of public opinion, in ways which the government could not have predicted.

The defeat of the Great Strike was of crucial importance to the Thatcherite cause, both in the conduct of coal policy, and on the wider political scene. Nevertheless, this was, in many respects, a lucky victory, which owed less to careful pre-planning by government, and more to the folly of the NUM leadership.

The Aftermath of the Great Strike: A Policy of Caution

So far as coal policy was concerned, Margaret Thatcher combined implacability of long-term aims with considerable circumspection in policy implementation: a combination which was to play a considerable part in the achievement of the government's objectives. Above all, she was concerned to consolidate the defeat of the Great Strike, and to prevent any resurgence of NUM power (however improbable such an outcome might have appeared in retrospect). For this reason, after the end of the strike, coal stocks at power stations were rapidly rebuilt to high levels. But this caution also had to be reconciled with progress towards the aim of creating a commercially viable coal industry. This involved finding ways of establishing a rate of colliery closures and manpower run-down which was politically acceptable, while at the same time allowing the industry to make steady economic progress.

To this end large sums of public money were made available to fund redundancy payments of sufficient generosity to ensure that miners would be keen to take 'voluntary' redundancy (overriding any protests by the unions at local or national level); and to reimburse BC for the 'social costs' of manpower rundown, in ways so organised that BC management was not inhibited from closing pits or laying off men by the financial cost of doing so. In effect, the government had decided that it was worth spending £60,000 (1998 money values) of taxpayers' money to facilitate each job lost in the coal industry. There was little or no attempt to assess the balance between these direct public expenditure costs, and wider 'social' costs arising, in terms of regional unemployment and the effects on mining communities – although there was limited funding of British Coal Enterprise, designed to facilitate job-creation in coalfield areas.

Yet there was no sense in which the government had pre-determined targets for the rate or *ultimate* extent of the reduction in manpower. The quantum of redundancies was negotiated with BC on a year-by-year basis, in the light of what was achievable. As it happened, in the five years after the Great Strike, most of the reduction in manpower can be attributed to the doubling of productivity rather than reductions in output. To a considerable extent, this increase in productivity can be attributed to the defeat of the strike, which not only enabled over 100 low productivity pits to be closed, but also provided the opportunity for BC mining engineers progressively to exploit the potential of the new coalface technology and rapid roadway drivages which were becoming available before the strike. But this opportunity for large productivity improvements was reinforced by the application of progressive financial targets set by government, which required cost reductions to be secured against a background of falling prices, and the generous government funding to facilitate 'voluntary' redundancies. The outcome was that between the end of the Great Strike, and the fall of Margaret Thatcher, the colliery labour force was reduced by two-thirds without any effective protest or political difficulty, and to all intents and purposes, irreversibly.

The Thatcher government, in spite of its free market rhetoric, also proceeded with considerable caution in the way that it allowed the coal industry to be exposed to market forces. In the coking, general industrial and domestic residential markets (where the scope for direct government action was in any case limited), the government was content to let BC's sales be eroded through the operation of long-term trends in favour of other fuels and (in the case of coking coal) higher quality imports. However, in the power station sector, which made up some 80 per cent of BC's total sales, and where the government still controlled a nationalised electricity industry, the policy was one of careful management of coal demand up to 1990/91.

The attempt by the first Thatcher government to erode coal's position (and hence the power of the NUM) by initiating a large programme of PWR nuclear stations was an almost total failure: in spite of a decade of strong government advocacy, only one PWR was built (Sizewell B: which was not commissioned until 1994). However, there arose a new threat to coal's market share in power generation.

Quite fortuitously, within a year of the end of the Great Strike, world oil prices collapsed, ushering in a new era of low international energy prices. In spite of large increases in productivity, BC was not able to reduce deep-mined costs sufficiently to match the steadily falling price of imported power station coal, which by 1990 had halved in real terms. In such circumstances, a 'free market' would normally have resulted in a substantial increase in power station coal imports. Yet this did not happen. Instead, the Thatcher government oversaw a series of agreements between BC and the electricity generators which mitigated the potential effect of market forces in order to produce a managed, gentle fall in the volume of BC sales to power stations, in return for a reduction in prices averaging 5 per cent a year in real terms between 1985/6 and 1992/3. Neither the nationalised CEGB, nor (in 1989/90) the newly-privatised generators were allowed to exercise their freedom in such a way as to prejudice the orderly management of the coal industry's decline: that is, with a politically acceptable rate of manpower rundown, consistent with 'voluntary' redundancy, and the continued observance by BC of formal colliery closure procedures.

As international coal prices continued to fall, the effect of these government-brokered arrangements was to create a large hidden subsidy for BC coal (as represented by the difference between BC and imported prices), which was paid by electricity consumers rather than taxpayers. This had a number of advantages for government, in that these arrangements were largely invisible to the electorate, and fell outside the constraints of the PSBR. Nevertheless, the scale and duration of this hidden subsidy in BC's power station coal price was inadvertent. At the time of each of the major negotiations of 'understandings'/contracts between BC and the generators (in 1986, 1989 and 1993) there was an apparently reasonable expectation that by the final year of each arrangement, BC's prices would broadly have converged with international prices. But this expectation was frustrated by the way in which international coal prices continued to fall much farther than either BC or the government had anticipated. Thus, although BC was subjected to a substantial reduction in the prices it could charge (thereby maintaining the pressure to reduce colliery costs), the government-brokered arrangements protected BC sales volumes and prices from the full impact of market forces to a greater degree than had been envisaged.

Electricity Privatisation and the 'Dash for Gas'

Although the achievement of the Thatcherite political objectives for coal entailed an end to the high dependence of UK electricity generation on BC coal, in 1990, the year Margaret Thatcher fell from power, BC sales to power stations were almost the same as in 1978/79, the last year of the previous Labour government: BC still provided around 60 per cent of the total power station fuelling requirement, reflecting both the government's failure to mount a large nuclear programme (as it had hoped to do in the early 1980s), and its efforts to protect BC from large-scale imports of power station coal.

Nonetheless, at the time of electricity privatisation in 1989/90, it was widely acknowledged that the change in the status of the major generators and the reorganisation of the electricity industry could have profound effects on the future of the UK coal industry. Although there is no evidence that the government had in mind any particular quantum of displacement of UK power station coal, the general expectation at that time of both government and BC was that, once the transitional government-brokered contracts with the major generators expired in March 1993, the coal industry would face more intense competition from imported coal.

These expectations were dramatically overturned. Instead, there was a rapid move into gas-fired generation, using CCGTs (the so-called 'dash for gas'): so far as the government was concerned, an unexpected development in terms of its scale and speed. The 'dash for gas' was the result of a particularly intricate combination of factors: the liberalisation of the UK gas market led to increased availability of natural gas, which was able to find a profitable outlet in power generation (using the new combined-cycle gas turbine technology) as a result of the structure of the newly privatised electricity industry, and the way it was regulated. The outcome was a very rapid growth in the applications to build new gas-fired plant. The Major government readily gave the required consents, even though, by general acknowledgement, most of the new CCGTs had total *costs* (including capital charges on new capital) in excess of the avoidable *costs* at existing coal stations. If this was a market, it was a very distorted one.

The Major government was far less cautious in its approach to the rate of rundown of deep-mine output than the Thatcher administration had been. It was the Major government's decision to allow the 'dash for gas' to develop, and to permit the major generators rapidly to reduce their coal stocks in the interim, which halved BC's sales to power stations in England and Wales in the three years before its privatisation in December 1994, leading directly to a precipitate reduction of over 40m. tonnes in annual deep-mined output. The scale and speed of this reduction – which accounted for over half the total reduction of deep-mined output which occurred over the whole period between 1979 and 1997 – led to the political 'coal crisis' of 1992/93. In large measure, this 'crisis' was of the Major government's own making from which it finally extracted itself with considerable political skill, but only by a settlement in 1993, which (despite initial appearances) provided for the coal industry only very temporary respite rather than solutions. The 'dash for gas', which was to prove to be by far the largest single factor in the reduction in deep-mined output was the largely unintended outcome of gas liberalisation and of market distortions arising from electricity privatisation. But the Major government did nothing to stop it, or to mitigate the severe consequences for the deep-mine coal industry.

Coal Privatisation

The final element in the achievement of the political objectives for coal was the completion of the privatisation of BC in 1994. However, although this was an objective entirely in tune with the principles of Thatcherism, Margaret Thatcher herself had proceeded with considerable circumspection. While some tentative steps were taken to erode BC's monopoly position in UK coal production, no serious consideration was given to the full privatisation of BC while she was Prime Minister, notwithstanding Cecil Parkinson's 'historic pledge' in 1988 to achieve this 'ultimate privatisation'. It was not until the privatisation of the electricity industry had been completed that John Major's government began in 1991 seriously to prepare for coal privatisation. Although the decision to privatise the ESI before BC was influenced by factors other than coal policy, nevertheless, in the way things turned out, this order of doing things facilitated BC privatisation. The successful privatisation of BC required that its deep-mined output should be reduced to around a sustainable economic level *before* the remaining mines were taken over by the private sector. But there would have been considerable political difficulties in doing so *ahead* of manifest contraction of the coal market. The 'dash for gas' provided a strong and very visible (albeit largely unpredicted) imperative to reduce output quickly. The successful completion of BC privatisation also depended crucially on the endowment of favourable contracts with the major generators, and the assumption by government of the large liabilities arising from past coal operations. But BC also had a part to play by securing further large reductions in costs and man-power, and by achieving a balance between supply and demand in a rapidly falling market. The success of BC in carrying out these tasks made a crucial contribution to ensuring that the new private owners would be largely free of politically sensitive and costly problems of colliery closures and redundancies in the years immediately following privatisation.

However, in the case of this particular industry, privatisation was no panacea. The sharp improvements in performance which generally followed the privatisation of utilities were, in the case of coal, *preconditions* for, rather than benefits arising from privatisation. Most of the sales and operating profits of the deep mines in the three years after BC privatisation derived from the terms of the inherited contracts with the electricity generators, brokered by the Major government in 1993. The problems which were to arise on the expiry of these contracts in March 1998, were bequeathed to the New Labour government. As the politics of colliery closures re-emerged, it became very clear that privatisation had done little or nothing to create a *sustainably* viable deep-mined coal industry in the private sector. Removing what the Conservative government had called the 'constraints of public ownership' did not overcome the coal industry's fundamental problems; nor did privatisation take the politics out of coal (as Thatcherite theory would have indicated). In these particular respects, notwithstanding initial apparent success, the Conservative political agenda for coal was not fully achieved.

The Role of NCB/BC

While much of our narrative has concerned the objectives and policies of successive Conservative governments, the part played by the management of the industry cannot be ignored, even though the position of the NCB/BC as an organisation was difficult and ambiguous throughout.

Before the Great Strike in 1984/5, the NCB's effective monopoly position in fuel for power generation thereby also conferred a virtual monopoly on the NUM, and so created the opportunity for political confrontation with the government, if the Union's leadership wanted it (as increasingly they did) and if they could persuade their membership to follow. Margaret Thatcher saw the NCB itself as contributing to this fundamental problem. Referring to the arrival of Ian MacGregor as the new NCB Chairman in 1983, she says: 'Within the NCB itself, he often found himself surrounded by people who had made their careers in an atmosphere of appeasement and collaboration with the NUM' (Thatcher p.342). Nigel Lawson went even further: 'The managerial inadequacies of the CEGB ... were as nothing compared with those of the Coal Board. It was the archetypal public corporation, where genuine business management was largely unknown. The pits around the country were run on a joint basis by mining engineers who were unversed in management, and the local representatives of the NUM' ... (and) 'the greater cohesion of the NUM tended to give it the upper hand most of the time' ... 'Neither management nor unions were much interested in either the costs of production or the demand for coal'. Lawson also repeats the Thatcher view that Hobart House (NCB Headquarters) was riddled with NUM moles; and refers to the 'Derek and Joe Show' whereby Ezra 'had in practice accepted the extraordinary powerful role in the industry's affairs assumed by the NUM, which he relied on Gormley to control' (Lawson pp. 154–5). While there were some elements of truth in this diatribe, it was a gross caricature of the general state of the industry's management and showed no recognition of the circumstances under which it operated. Before the Great Strike, the NCB management had been faced with a dual dilemma: first, how to pursue a policy of major investment under 'Plan for Coal', designed to consolidate UK coal's perceived strategic importance in the era of OPEC power, without at the same time consolidating the power of the NUM; and second, how to make progress with eliminating surplus, uneconomic production capacity while seeking to keep the peace in the face of the annual cycle of large wage demands and strike threats. There is room for honest disagreement as to whether at all times the appropriate balance was struck between these various conflicting demands; but there is no denying the great difficulty of the task.

The defeat of the Great Strike in 1985, while largely removing the NUM's ability to impede management's efforts to increase productivity and close high-cost pits, also demolished much of the industry's protectionist defences. This created a different sort of dilemma for BC management: how far should the Corporation behave like any other business? There were two main difficulties here. First, there was a fundamental lack of clarity of objectives going back to the Coal Industry Nationalisation Act of 1946 (drawn up in very different circumstances), particularly the ambiguous status of the statutory financial targets, which were viewed as constraints rather than primary objectives. Although the Conservative government's promulgation of a series of

'Chairman's Objectives' (beginning in 1983) sought to place greater emphasis on profitability, it was unlikely that *in fact* the government wished BC to follow profit *maximisation* by eliminating systematic cross-subsidies, since such a course would have involved an even greater rate of manpower rundown than the large programme which both BC and the government were having to cope with in any case. Second, BC's difficulty was compounded by its lack of independence arising, not only from its constitutional position as a nationalised industry (being subject to covert political interference as well as overt political accountability), but also from the industry's dependence on government finance, and on government favours in the conduct of the crucial relationship between the coal industry and the major electricity generators. BC's position as a nationalised industry effectively precluded radical independent changes in the direction of corporate strategy in response to radically different circumstances facing the industry. Fundamentally, NCB/BC were bound to act as government agents, but agents who could be disowned if difficulties or political embarrassments occurred.

In all the circumstances, it is not surprising that progress with the cultural revolution necessary to turn BC from an institution into a business was for some time sporadic and incomplete. Yet to dwell too much on this may obscure the very real contribution which BC management made to the delivery of the government's political objectives after the Great Strike, by the expertise brought to bear in securing a five-fold increase in productivity, with associated cost reductions, and managing the process of decline. The only serious problem in carrying out that task occurred at the time of the coal crisis of October 1992, which was of the government's, not BC's making.

There are some who would argue that if BC had behaved more like an efficient, private enterprise, and had reduced costs faster, much of the precipitate fall in deep-mined output which took place between 1991 and 1994 could have been avoided. While it is always difficult wholly to disprove such hypotheses, several points can be made in BC's defence. First, the attainment of a nearly 60 per cent reduction in average colliery operating costs over the decade between the Great Strike and BC privatisation was, by any standards, a very significant achievement. Second, assessments of potential improvements made by external mining consultants, appointed by government in 1991, did not materially differ from those put forward by BC itself. Third, and most important, the very rapid fall in deep-mined output in the early 1990s was due to the 'dash for gas' and (to a lesser extent) the improving performance of nuclear power stations, together with the rapid reduction of the generators' coal stocks. None of this would have been materially different if BC costs had been somewhat lower.

All things considered, we can reasonably maintain that the professional competence of the nationalised industry's management in increasing productivity while managing the huge decline in manpower, and in preparing the way for coal privatisation, deserves to be added to the list of important factors which enabled the Thatcherite political agenda for coal to be so completely achieved by the time that BC was privatised.

Adverse Economic Fundamentals

Going back to the beginning: there is absolutely no evidence that the first Thatcher government either intended, or even expected, that UK deep-mined output would fall from over 100m tonnes a year in the early 1980s to as low as 30m. tonnes in the mid-1990s. Even after the defeat of the Great Strike, the government's emphasis was not (as some have maintained) on 'destroying' the coal industry, but on turning it into a sustainable commercial operation; and while there was clear recognition that this was likely to involve contraction (particularly of manpower), there was no suggestion that such contraction should continue *beyond* the point which could be justified by economic criteria. Unfortunately for BC, the fundamental economic trends affecting UK coal continued to be adverse, so that what might be regarded as an 'economic size' for the industry became an ever downward-moving target to an extent not envisaged either by BC or the government.

As we have noted, the long-term erosion of UK coal's position in the coking, general industrial and domestic residential markets, continued unabated, as a result of industrial restructuring, technological change, and social preference. Over the whole period 1979 to 1997, the reduction in sales to those markets was to amount to nearly 30m. tonnes, equivalent to some 40 per cent of the total reduction in deep-mined output between those years. But it was in the power generation sector, which represented around 80 per cent of BC's markets, that the main economic and public policy issues arose.

Between the early 1980s and 1994 (when BC was privatised), the market value of UK deep-mined coal, as measured by the delivered sterling price of competing internationally-traded power station coal, fell by some two-thirds in real terms. Although BC managed to reduce average colliery operating costs by approaching 60 per cent, these cost reductions could not have been achieved without the closure of many high-cost pits, and were nevertheless insufficient to make economic the majority of the remaining deep-mined output supplied to UK power stations. (See Table 8.1). Further important conclusions can be drawn. First, it was very evident that the marginal markets for UK power station coal – namely UK power stations distant from the coal fields or exports – were uneconomic by a very wide margin. Second, even at inland power stations, which remained BC's largest and most favourable market, there remained a significant (albeit diminishing) negative margin against *average* UK deep-mine costs, notwithstanding the very large reduction in deep-mined output which took place, particularly after 1991. Given that the pits which closed were generally of higher cost than those that remained, this strongly suggests that

Table 8.1: BC Deep-mined Operating Costs against Imported Coal Prices. £/GJ at 1998 Money Values

	Average Deep-mined Operating costs	Competitive Value against Imports	
		Inland Power Stations	Thames-side Power Stations
1981/2	3.10	2.95	2.55
1982/3	3.14	3.17	2.77
1983/4	STRIKE AFFECTED		
1984/5	STRIKE AFFECTED		
1985/6	2.98 *	2.34	1.94
1986/7	2.61	2.13	1.73
1987/8	2.52	1.51	1.11
1988/9	2.24	1.54	1.14
1989/90	2.30	1.72	1.32
1990/1	2.07	1.42	1.02
1991/2	1.89	1.42	1.02
1992/3	1.74	1.55	1.15
1993/4	1.47	1.36	0.96
1994/5 (9 months)	1.38	1.24	0.84

Notes:

* Excluding release of provision for strike loss

1998 Money Values calculated by use of GDP Deflator (Market Prices)

'Competitive Value' represents BC pithead price equating with the delivered price of imports, derived from IEA Data on EU Power Stations Coal Imports from Third Countries by *calendar* years, adjusted for estimated transport differentials of 20p/GJ in favour of BC at inland power stations, and 20p/GJ against BC at Thames-side power stations. Although these transport differentials are broad averages, which would vary significantly in the case of individual collieries and power stations, they give a reasonable view of the overall competitive position of UK deep-mines against imported steam coal. (Depending on location, the value of export sales were broadly similar to those of sales to Thames-side power stations).(See Note 5 to Table 7.2 Cm. 2235) even if there had been no significant move into gas-fired generation, and all the competition in the 1990s had come from imported coal (which would probably have been the case if there had been an undistorted free market in power station fuelling), the *economic* size of the deep-mined industry – that is, the level of output which could be produced without systematic internal cross-subsidy, at costs below competitive prices – would if anything have been less than the output levels which actually obtained.

The adverse trends in the market value of UK deep-mined coal relative to its costs were also reflected in the industry's capital expenditure, and hence its future potential. Like any extractive industry, capital expenditure on new and replacement coal production capacity was a pre-requisite if long-term decline in deep-mine output was to be avoided. In 1979, the new Thatcher government inherited the 'Plan for Coal' which had originally been initiated by the Heath government during the 1973/4 OPEC oil crisis, and implemented by the Labour governments of 1974–9, when most of the major investment decisions were taken. The momentum of this programme was maintained during the early 1980s, at a level sufficient to maintain deep-mine capacity for a few years ahead. However, thereafter capital expenditure in real terms, expressed as a rate per tonne of current deep-mined output, fell progressively, then rapidly in the years before privatisation, and still further after privatisation, by which time it was only a seventh of the level before the Great Strike, and a long way below what was needed to prevent future output decline (See Table 8.2).

Table 8.2:   Trends in Coal Industry Capital Expenditure Relative to Deep-Mined Output

	Average Annual Capital Expenditure		Average Annual Deep-Mined Output	Average Cap.Ex per tonne of Deep-Mined Output
	Money of Day £m.	1998 Money £m.	M.tonnes	1998 £/tonne
1979/80–1983/4	715	1548	104.7	14.8
1985/6–1989/90	602	929	83.9	11.1
1990/1–1993/4	250	301	62.0	4.9
1995–1999*	51	52	24.0	2.2

*   RJB Mining (previous years NCB/BC)

Notes: NCB/BC figures based on annual additions to assets (nearly all of which was on deep mines)
GDB Deflator used
1984/5 (Great Strike) and 1994/5 (9 months before privatisation) omitted.

From the mid-1980s there had been a growing realisation that, because the economics of new and replacement deep-mined capacity had become very poor, economic UK deep-mined reserves would be largely confined to the best seams at the surviving mines which could be accessed without major capital expenditure. Moreover, the mining technology using heavy-duty face equipment, which had played such an important part in the very large increases in productivity after the Great Strike, was both capital intensive and inflexible in dealing with variable geology, so that high productivity entailed increasingly selective working of the available reserves. All this meant that the deep-mine industry was bound to decline. The fallacy of '300 years of coal' was exposed, and economic coal reserves, which continued to decline rapidly after privatisation, were shown to be much more limited than those of North Sea oil and gas. (See Parker: 'Energy Exploration and Exploitation' Vol. 15, No.1: (1997) 'UK Coal Reserves in Perspective' and Vol. 17, No. 6 (1999) 'UK Deep Mined Coal Reserves: Further Declines Inevitable') (Also IMCL Report in Chapter 7).

By the end of our study although the major capital expenditure involved in creating the surviving mines (and the power-stations to which they sent most of their output) had already been sunk, the remaining deep-mine industry was still unable to avoid further progressive decline.

'Going with the Grain'

Looking with a somewhat wider perspective, it could be seen that the substantial decline of UK deep-mined output and its labour force, and hence of the influence of the mining unions, went 'with the grain' of events. The fundamental economics of the industry indicated that stability was not a realistic option: the only real question was the rate and extent of decline. But the government could contemplate this with equanimity, since the collapse of the power of OPEC in 1986 (not long after the collapse in the power of the NUM), and new perceptions that oil, gas and imported coal would remain abundant, removed much of the force of the earlier arguments in favour of UK coal on grounds of long-term strategic energy security. Moreover, growing environmental concerns – first acid rain and then global warming – were seen as damaging to UK coal. Although in the period of our study, little or no *direct* loss of market for deep-mined coal can be attributed to environmental regulation, the impact of *future* environmental constraints was a significant factor in later years in discouraging future coal-fired generation (particularly as 'clean coal' technology was substantially more expensive than gas-fired plant), and creating expectations of further reductions in the market for coal.

Policy on coal also went 'with the grain' of the Conservative government's stance on public policy generally, with the emphasis on reducing the size of the public sector, furthering competition and market liberalisation, and reducing the power of the trades unions. This minimised any potential conflicts over coal policy within the government or within the Conservative Party (except during the 'coal crisis' of October 1992, when the government's management skills temporarily deserted it). But there is another and obvious point. The Conservative Party was in power continuously from 1979 to 1997, and faced little effective political opposition to its coal policies (except briefly during the exceptional circumstances of the 'coal crisis' of 1992/3), even though the Labour Party was totally opposed to the government's policy on coal.

An Alternative Political Scenario

However, the total absence of a bipartisan approach to the coal industry's problems meant that the result of successive general elections had a major effect on the ultimate size and condition of the industry. If Labour had won the 1983 election, it is inconceivable that there would have been a major defeat of the NUM at the hands of the government. The huge reduction in the workforce and number of pits which took place during the second Thatcher government would not have occurred. The NCB/BC would have been condemned to solve the problem of surplus, high-cost capacity by messy gradualism. If Labour had won the 1987 election, electricity privatisation would not have occurred; and the 'dash for gas' (a large part of which was given government approval before the 1992 election) would not have been launched. If Labour had won the 1992 election, it is difficult to believe that measures would not have been taken to slow the rate at which deep-mine output was falling (for example, by not allowing the generators to run down their coal stocks so rapidly, and by stopping further CCGT building); and the final step in Conservative policy – the privatisation of BC itself – would not have taken place.

We can, therefore, consider an alternative scenario, on the basis of a continuously 'pro-coal' Old Labour government from 1983 onwards. Under this scenario, the government would have retained control over a nationalised electricity industry: and would have broadly stabilised consumption of UK coal in power stations at not less than 70m. tonnes a year, by disallowing the use of gas, not proceeding with further nuclear stations beyond those already authorised, by restricting net imports of electricity over the French interconnector, and by limiting coal imports for power generation. Such measures would have been supported by a series of government sponsored long-term arrangements or contracts between the coal industry and the ESI, together with a programme of FGD installations, and some 'clean coal' stations, sufficient to protect the coal industry from EU environmental regulation of SO_2 emissions. On the other hand, this scenario assumes that the government would have made no serious attempt to mitigate the long-term downward trends in coking, industrial and domestic residential markets for UK coal; and that opencast output would be broadly maintained at about 15m. tonnes to complement deep-mined output and provide a financial cross-subsidy.

Under this scenario, there would still have been a market for some 65m. tonnes of deep-mined output in 1997 (compared with an actual level of 30m. tonnes in that year). Over the period 1982–97, this would have represented an annual average output erosion of nearly 3m. tonnes per annum: most of this being consistent with the likely reduction in colliery capacity due to natural exhaustion, on the not unreasonable assumption that this 'pro-coal' government would nevertheless not have embarked on a major investment programme of new and replacement capacity (a new 'Plan for Coal'), given the huge changes in international energy markets which followed the collapse of oil prices in 1986, and other calls on public expenditure.

But the other side of this alternative scenario is that it would most likely have been associated with very much lower productivity growth and continuing inhibitions on the closure of high cost pits, except where they had effectively exhausted their reserves. After the Conservative government's retreat over closures in February 1981, the NCB advised that the long-run productivity growth rate would be about 3 per cent per annum. On this (admittedly speculative) basis, and with deep-mined output falling progressively to about 65m. tonnes, the NCB/BC colliery labour force would have been reduced to about 85,000. (Table 8.3)

Thus, one might conclude very broadly that, even with a non-Thatcherite, 'pro-coal' Old Labour government from 1983 to 1997, deep-mined output would probably have fallen by 40m. tonnes (mainly as a result of decline in non-electricity markets), and manpower would have fallen by about 120,000.

Nevertheless, the difference in outcome would have been much

Table 8.3: Alternative Political Scenario: Hypothetical Projections and Actual
Outcomes

	1982/3 Actual	1997 Actual	1997 'Pro-coal' scenario
Deep-mined output (mt)	105	30	65
Average Colliery Manpower ('000)	208	12	85

more than just a question of size: the scenario of a 'pro-coal'
government would have delivered a very different sort of industry. A
productivity growth of only 3 per cent a year (assuming that is what
would have happened) associated with real wage increases would have
yielded only modest reductions in operating costs. Thus, by the mid-
1990s UK deep-mine costs would have been around three times the
equivalent international prices, with the industry being sustained only
by huge subsidies either from the taxpayer or the electricity consumer.
(The economic position of the industry would have been even worse
if there had been attempts to mitigate the manpower losses by short-
time working). Moreover, with the ESI still some 60 per cent dependent
on UK coal, the problem of the excessive power of the NUM would
have remained unresolved.

Consideration of a 'pro-coal' government scenario raises also a
comparison with the position of deep-mining in Germany. (Table 8.4)
Productivity growth, as measured by output per man-year, was little
more than 1 per cent a year, and by 1997 was only a quarter of the
UK productivity level.

Table 8.4: Deep-mining Industry in Germany. 1985 and 1997

	1985	1997
Output (m. tonnes)	82	47
Manpower	166	78
Output per man year	492	593

Source: IEA *Coal Information* 1998

Unlike Conservative government policy in the UK, the main preoccupation of German government policy on coal was to provide the
mechanisms of subsidy and other forms of support sufficient to allow
the industry to contract gradually at a rate consistent with broad
agreement between the government, coal industry and mining unions.

The politics of coal in Germany was greatly influenced by the checks and balances in the federal structure of German politics, which sustained the influence of the SPD-dominated coal-producing North Rhine Westfalia, whereas in the UK, with its highly centralised governmental system, there were no such regional checks of sufficient influence to mitigate the rigour of Thatcherite policy. Further, in contrast to the strong political dimension to coal policy in the UK, influenced by the confrontational stance taken by Arthur Scargill and his associates in the NUM, the German Mining Unions had been models of moderation, and willing participants in the joint consultation and consensus approach adopted within the German coal industry, and in its relationship with the Federal Government. As a result, the German industry was heavily protected by huge subsidies and subject to only gradual reductions in output and manpower, while the UK industry, with costs at less than half the German level was rapidly contracted. The European Commission seemed powerless to deal with this 'Anglo-German anomaly'. (See Parker: 'The Politics of Coal's Decline: The Industry in Western Europe' RIIA, 1994)

'Counterfactuals' are always hazardous. However, both our speculative hypothesis on the effect of a 'pro-coal' Old Labour government between 1983 and 1997, and actual experience in Germany, suggest that, although significant contraction of deep-mined output and manpower would have occurred in any case, the rate of contraction would have been much less marked than that which occurred under successive Thatcherite Conservative administrations; (although the economic position of the surviving industry would have been very much worse).

Finale

We have seen that there was, from the earliest years of Margaret Thatcher's administration, a 'political agenda' for the coal industry, based on Thatcherite principles of breaking the power of over-mighty trades unions, promoting competition and free markets, and with a marked preference for private, rather than public ownership. In the case of the coal industry, Thatcherite policy proved to have considerable internal coherence and exerted a crucial influence in breaking the power of the NUM, reducing deep-mined output and manpower, and moving the industry from the public to the private sector, in ways which would not have occurred under plausible alternative governments.

We have also seen how the actual process by which the original Thatcherite political objectives for coal were achieved, was characterised by a combination of the government's ultimate clarity of purpose and its political skill (except in the first years of the Major government) – and made possible by the fact that the Conservatives were in power for four Parliaments: a period long enough to allow the desired outcome to occur. But the outcome was also the result of chance, unexpected consequences, and economic trends that could not have been foreseen when Margaret Thatcher came to power in 1979. Thus the Conservative governments were attended by good fortune, being greatly assisted by the folly of the NUM leadership (which was a decisive factor in enabling the government to defeat the Great Strike), the professionalism of BC in managing decline, the unforeseen way in which electricity privatisation was followed by the 'dash for gas', and the unpredicted severity of the progressive economic and other external pressures on the UK coal industry. The profound changes in world energy markets, in particular after the collapse of OPEC oil prices in 1986, led to the real *value* per tonne of UK coal output falling by 1997 to only a third of the level which obtained before the Great Strike of 1984/85. In spite of large increases in productivity and the closure of many high-cost pits, the economic pressures to reduce deep-mined output continued unremittingly; and, given the very unfavourable economics of replacement investment in this old extractive industry, on-going decline became inevitable. *Both the economic fundamentals and the government's objectives pointed in the same direction.*

This brings us to the real nature and significance of the Thatcherite political agenda for coal. This was never a detailed pre-conceived action plan; but rather a set of inter-related principles, strongly held over a long period, which could be brought to bear on ever-changing circumstances. The Thatcher and Major governments saw the emerging downward pressures on deep-mined output and manpower, not as forces to be resisted by government action (as the Germans did and Old Labour would have done), but as developments to be accepted, even exploited, subject only to the dictates of political prudence.

As to the final outcome, the irreversible demolition of the power of the NUM represented an important change in the industrial and political landscape of Britain. But neither the massive 'down-sizing', nor what the Conservatives hoped would be the magic wand of coal privatisation, was able to create, even on a much reduced scale, a *sustainably* viable deep-mined industry. In 1997, the incoming Blair government adopted for the energy sector the Thatcherite framework of private companies operating in competitive market, only to find that the coal question sat on the fault-line between Old and New Labour, and that measures designed to 'save the coal industry' could not prevent the continuing decline of the deep mines. And so, by the turn of the century – some twenty years after Margaret Thatcher became Prime Minister – what remained of the once-great UK deep-mine coal industry entered upon, in a somewhat precarious state, the last stage in its long history.

SOURCES AND BIBLIOGRAPHY

Industry Sources

The affairs of the UK coal industry over the last two decades have given rise to an immense amount of written material. Our task has been selection rather than discovery of the relevant documentation (particularly as our study focuses on the interaction of politics, economics and the unexpected in determining the industry's fortunes since 1979, rather than providing a fully-comprehensive description of all the industry's activities over that period.

A great deal of the material is already in the public domain, from evidence presented to public enquiries, government reviews, and Select Committees. Much can be gleaned from the series of Annual Reports and Accounts of the National Coal Board/British Coal Corporation down to 1994/5, when the mining assets were privatised. (The National Coal Board was renamed the British Coal Corporation in March 1987.) For the period since privatisation, the major industry sources are the Annual Reports and Accounts of RJB Mining PLC, Mining (Scotland) Limited, and Celtic Energy Limited (Celtic Group Holdings Limited).

In addition, the author has consulted the administrative papers and records of the National Coal Board/British Coal Corporation (with the permission of the Corporation). Although NCB/BCC papers have the status of Public Records, many of those consulted in this study will not be made public for many years. It would, therefore, be inappropriate and often impracticable to give full documentary references to all the NCB/BCC papers consulted. The written record has also been supplemented by the author's own recollections from his service at BCC (until late 1991), and by the recollections of some of his former close colleagues.

House of Commons (Select) Committees: Reports

(i) Energy Committee
Pit Closures. HC 135 December 1982
The Coal Industry. HC 165 January 1987

The Government's Response to the Committee's Report on the Coal
   Industry. HC 387 May 1987
The Structure, Regulation and Economic Consequences of Electricity
   Supply in the Private Sector. HC 307 July 1988
Energy Policy Implications of the Greenhouse Effect. HC 192 July
   1989
The Flue Gas Desulphurisation Programme. HC 371 June 1990
Government Observations on Committee's Report on the Flue Gas
   Desulphurisation Programme. HC 662 October 1990
Consequences of Electricity Privatisation. HC 113 February 1992

(ii)   Trade and Industry Committee
British Energy Policy and the Market for Coal. HC 237 January 1993
   Coal (2 volumes). HC 404 March 1998

(iii) Employment Committee
Employment Consequences of British Coal's Proposed Pit Closures.
   HC 263 January 1993

Reports by Government and Other Public Bodies

(i)   Department of Energy
Coal Industry Examination: Interim and Final Reports. June and
   October 1974
Sizewell B Public Enquiry: Report by Sir Frank Layfield. 1987
Privatising Electricity: The Government's Proposals for the Privatisation
   of the Electricity Supply Industry in England and Wales (White
   Paper). Cm. 322. 1988 (Proposals for Scotland were contained in
   Cm. 327. 1988)

(ii) Department of Trade and Industry
The Prospects for Coal: Conclusions of the Government's Coal Review
   (White Paper). Cm. 2235. March 1993
The Energy Report Vol.1. 'Markets in Transition'. 1994
Conclusions of the Review of Energy Sources for Power Generation
   (White Paper). Cm. 4071. October 1998
Energy Paper 67: Cleaner Coal Technologies. April 1999
UK Energy Sector Indicators 1999
The Energy Report 1999

(iii) Department of Trade and Industry: Consultants' Reports
John T. Boyd Company. Independent Analysis. 21 Closure Review
    Collieries, British Coal Corporation. Report No. 2265.6. January
    1993
Prospects for Coal Production in England, Scotland and Wales. Report
    by International Mining Consultants Ltd. March 1999

(iv) Monopolies and Mergers Commission
National Coal Board: A Report on the Efficiency and Costs in the
    Development, Production and Supply of Coal by the NCB. Cmnd
    8920. June 1983 (2 volumes)
A Report on the Investment Programme of the British Coal
    Corporation. Cm. 550. January 1989
National Power PLC and Southern Electric PLC. Cm. 3230. 1996

(v)   Commission on Energy and the Environment
Coal and the Environment. (HMSO) 1981

(vi) Department of the Environment, Transport and the Regions
Minerals Planning Guidance: Coal Mining and Colliery Spoil Disposal
    (MPG3 Revised). March 1999

(vii) Her Majesty's Inspectorate of Pollution (HMIP)
News Release HM 396: Further Substantial Reductions in Air Pollution
    from Power Stations in the Electricity Supply Industry. March 1996

(viii) Office of Electricity Regulation (OFFER)
Review of Economic Purchasing. February 1993

(ix) The Coal Authority
Annual Report and Accounts 1994/5 to 1999/00

Other Published Sources

Ashworth, W. *The History of the British Coal Industry*. Volume 5: 1946–
    1982 *The Nationalised Industry*, OUP 1986
Coalfield Communities Campaign
        A Fair Deal for Coal – A Fair Deal for Britain. January 1997
        A Market for Coal. January 1998
Hattersley, R. *Fifty Years On*, Little, Brown & Co. 1997
Heath, Edward *The Course of My Life*, Hodder and Stoughton. 1998

Heseltine, Michael *Life in the Jungle*, Hodder and Stoughton. 2000

Lawson, Nigel *The View from Number 11*, Bantam Press. 1992

Major, John *The Autobiography*, Harper Collins. 1999

Ottey, Roy *The Strike*, Sidgwick and Jackson. 1985

Parker, M.J. *The Politics of Coal's Decline: The Industry in Western Europe*, Royal Institute of International Affairs. 1994

> *Energy Exploration and Exploitation* (Multi-Science Publishing)
> 'UK Coal Reserves in Perspective', Vol. 15. No.1. (1997)
> 'UK Deep Mined Coal Reserves: Further Declines Inevitable', Vol.17. No. 6. (1999)

Parker, M.J. and Surrey, A.J. *UK Gas Policy: Regulated Monopoly or Managed Competition?* SPRU. 1994

Parkinson, Cecil *Right at the Centre*, Weidenfeld and Nicolson. 1992

Ridley, Nicholas *My Style of Government*, Hutchinson. 1991

Routledge, Paul *Scargill*, Harper Collins, 1993

Smith, Ned *The 1984 Miners' Strike: The Actual Account* (Private Publication. 1997)

Thatcher, Margaret *The Downing Street Years*, Harper Collins. 1993

Yergin, Daniel *The Prize*, Simon and Schuster. 1991

Miscellaneous

Since 1992, the monthly journal *Coal UK* (McCloskey Energy Publishing) has provided a very useful commentary upon events.

For UK Statistics (other than industry sources), the main sources have been the annual *Digest of United Kingdom Energy Statistics*, and the monthly *Energy Trends* (Department of Trade and Industry).

For international coal statistics, the principal source has been the annual *Coal Information* reports by the International Energy Agency (OECD).

INDEX